Praise for *Chips, Clones, and Living Beyond 100*

"A great, useful, and timely book, which gives much insight into biomedicine and takes away the fears that many may have. A must for lay people who want to know how far bioscience has come and where it will go, as well as for scientists in other fields."

> **—Dr. Hans Vemer**, Schering-Plough; formerly President of Organon International Schering-Plough

"This book needed to be written. Financiers, executives, and policy advisors will benefit greatly from this broad overview of the most important developments in the bewildering field of biotechnology. The combination of advanced technology insights and managerial guidance about managing extreme uncertainty make this book truly unique."

> **—Donald Kalff, Ph.D.**, Biotech Entrepreneur and Venture Capitalist; former member KLM Executive Board

"The authors offer a splendid 'tour of the horizon' that considers the breakthrough consequences of current and future developments in the biosciences. Readers can profit in many ways from this book, whether in business, education, government, or at home."

> **—Professor George S. Day**, The Wharton School, University of Pennsylvania

"As physics was to the 20th century, the biomedical sciences will be to the 21st. The explosion of new knowledge in areas such as genetics, neuroscience, stem cell biology, artificial organs, and regenerative medicine promises to improve our lives, create new business opportunities, and permit us to control many diseases while raising important challenges for our social, ethical, and economic thinking. The Schoemakers have given us an incredibly useful book to stimulate that thinking and to shed light on what the next two decades might and should hold in store. One could not ask for a better guidebook to an exciting, if challenging, future."

> **—Professor Arthur L. Caplan**, Director of the Center for Bioethics and Hart Professor, University of Pennsylvania

"The growth of the biosciences is a global phenomenon. Singapore is betting on it as its next growth engine via Biopolis. This book offers a rich roadmap for the next wave of major biomedical innovations."

—**Marvin Ng**, DN Venture Partners, Singapore

"A stimulating and exciting look at how we got to the present state of health care and where we can potentially go. This is a must read for anyone interested or working in the health care arena, from the layperson to the entrepreneur to the corporate executive. This book is a unique perspective and a great read."

—**David Lester, Ph.D.**, President, ITHW Inc.; formerly Director, Human Health Technologies, Pfizer Inc.

"What a fascinating book! The authors have really mastered all the aspects (social, human, scientific, and business) of the biomedical revolution that is taking place this century. Awesome."

—**Giancarlo Barolat**, M.D. Board Certified by the American and Italian Board of Neurological Surgery; Director, Barolat Neuroscience, Presbyterian St. Luke Medical Center, Denver; formerly Professor of Neurosurgery at Thomas Jefferson University, Philadelphia

"Wonderfully comprehensive, yet still digestible for non-scientists. Wish I had this book when we examined some health care innovation opportunities at GE; it would have provided a great foundation for the team."

—**Patia McGrath**, Global Director - Innovation and Strategic Connections, Corporate Marketing, General Electric

Chips, Clones, and Living Beyond 100

How Far Will the Biosciences Take Us?

Paul J. H. Schoemaker, Ph.D.
Joyce A. Schoemaker, Ph.D.

Vice President, Publisher: Tim Moore
Associate Publisher and Director of Marketing: Amy Neidlinger
Editorial Assistant: Pamela Boland
Development Editor: Russ Hall
Operations Manager: Gina Kanouse
Digital Marketing Manager: Julie Phifer
Publicity Manager: Laura Czaja
Assistant Marketing Manager: Megan Colvin
Cover Designer: Stauber Design Studio
Design Manager: Sandra Schroeder
Managing Editor: Kristy Hart
Senior Project Editor: Lori Lyons
Copy Editor: Gayle Johnson
Proofreader: Kathy Ruiz
Senior Indexer: Cheryl Lenser
Senior Compositor: Jake McFarland
Manufacturing Buyer: Dan Uhrig

FT Press Science offers excellent discounts on this book when ordered in quantity for bulk purchases or special sales. For more information, please contact U.S. Corporate and Government Sales, 1-800-382-3419, corpsales@pearsontechgroup.com. For sales outside the U.S., please contact International Sales at international@pearson.com.

Printed in the United States of America

First Printing September 2009

ISBN-10: 0-13-715385-6
ISBN-13: 978-0-13-715385-5

Pearson Education LTD.
Pearson Education Australia PTY, Limited.
Pearson Education Singapore, Pte. Ltd.
Pearson Education North Asia, Ltd.
Pearson Education Canada, Ltd.
Pearson Educatión de Mexico, S.A. de C.V.
Pearson Education—Japan
Pearson Education Malaysia, Pte. Ltd.

Library of Congress Cataloging-in-Publication Data

Schoemaker, Paul J. H.

  Chips, clones, and living beyond 100 : how far will the biosciences take us? /
Paul J. H. Schoemaker, Joyce A. Schoemaker.

    p. cm.

  ISBN 978-0-13-715385-5 (hardback : alk. paper)   1. Biotechnology--Forecasting.
2. Biotechnology--Forecasting.   3. Medical sciences--Forecasting. I. Schoemaker, Joyce A.
II. Title.

  TP248.215.S36 2009

  660.6--dc22

                                                                    2009023679

Dedication

We dedicate this book to the memory of Hubert J. P. Schoemaker, our beloved brother and brother-in-law, who was a genuine pioneer of the new biosciences. After finishing his Ph.D. in biochemistry at MIT in 1977 and working briefly for Corning Biomedical, Hubert joined a small group of scientists and entrepreneurs in 1979 to found one of America's first biotech companies, Centocor Inc. It became the leader in the exciting new field of monoclonal antibody technology, moving gradually from diagnostics into therapeutics.

The company could be the poster child for the roller-coaster ride typical of many biotech companies. Its stock dropped from $60 per share to less than $6 per share when the FDA failed to approve its promising drug Centoxin for septic shock in 1992. Through alliances with Eli Lilly and other established pharma companies, Centocor rose from the ashes with an anticlotting medicine for coronary disease (ReoPro), followed by a major blockbuster drug for Crohn's disease, rheumatoid arthritis, and other autoimmune diseases (Remicade). Hubert lived through these ups and down as president, CEO, and chairman. He oversaw the final sale of the company in 1999 to Johnson & Johnson for nearly $5 billion, after it had regained its lofty stock price of around $60 per share.

Tragically, Hubert died from brain cancer in 2006 at age 55. Even while battling this unfair and harsh illness, he founded and ran a new company devoted to adult stem cells. He continued to operate at the frontiers of science and business until his untimely death. Hubert's courage, leadership, philanthropy, humanity, and perseverance in the face of great adversity will remain an inspiration for all those who knew him personally.

Contents at a Glance

Contents

Foreword

This book takes the reader on a virtual journey into the future, to imagine the consequences of the convergence of life science breakthroughs and information technology advances. We have already seen the impact of this convergence on lengthened life expectancies, thanks to personalized medicine, remote diagnostics, biosensors, tissue engineering, and precision imaging that reveals the disease process. Yet these are still early days. Paul and Joyce Schoemaker provide a splendid "tour of the horizon" that considers the breakthrough consequences of current and future developments.

For each of the possibilities that can be envisioned, there are profound personal (What quality of life will I have?), social (How will family relationships change?), regulatory and ethical (Who decides what should be encouraged?), economic (Can society afford a rapidly aging population?), and business (Who will capture the economic value that is created?) implications. As with any disruptive technology wave, clear business winners and losers will emerge. But especially in this case, we expect that the real beneficiaries will be individuals and their families.

The path of the biosciences revolution will be strewn with scientific dead ends, ethical dilemmas, failed experiments, public anxiety, and recurring budget crises—but also with as-yet-unknown scientific breakthroughs. We don't know how these uncertainties will play out in the next 10 to 15 years, but the range of possibilities can be captured by the scenarios introduced in Chapter 7, "Wildcards for the Future."

This is where I enter the picture as Co-director of the Mack Center for Technological Innovation at the Wharton School, where Paul serves as Research Director. Our Mack Center was the sponsor of the original scenario study on the "Future of the Biosciences" in 2005. This was one of our most successful initiatives: We became a credible and informed player in the ongoing debate about the role of emerging technologies in the biosciences, building on our decade-long inquiry into the challenges of managing emerging technologies across a wide range of industries. In 2003, we concluded that the lessons from this broad inquiry needed to be tested more deeply in specific industries, and we chose the biosciences. This is a complex sector where development processes take years, success is heavily influenced by regulators, and all the usual risks of innovation are magnified. Using our deep and wide relationships with industry leaders, our center launched the Biosciences Crossroads Initiative to dive more deeply into this important and dynamic sector.

Paul and Joyce have been an integral part of the intellectual journey that we began 15 years ago with the Mack Center. I feel privileged to write this foreword as a token of my gratitude for their friendship and my appreciation for all that I have learned from them. The best part of our journey is that it is never finished. There will always remain deep questions, exciting new possibilities, and important lessons from which we can generalize. This book takes stock of where we are now in understanding the complex dynamics of the biosciences, while also emphasizing the important process of continual learning and discovery. Readers can profit in many ways from this book, whether in business, in education, in government, or at home.

—George Day, Boisi Professor
Professor of Marketing
The Wharton School, University of Pennsylvania

Preface

The 20th century might have belonged to physics, but the 21st will be about the biosciences and their impact on human health and well-being. This diverse collection of technologies is changing our world in dramatic ways, with the potential for even greater impact in the years ahead. In addition to improving human health, many other industries will be impacted by the biosciences—agriculture, forestry, food processing, and legal services, among others. The influence is also being felt in energy, cosmetics, supermarkets, the military, household appliances, information services, homeland security, and so on.

Defining bioscience and biomedicine

Bioscience consists of existing life science disciplines, such as biology and biochemistry, as well as related technologies, especially genomics, proteomics, bioinformatics, stem cell therapies, cloning, and mono-clonal antibody therapies. Adjacent and overlapping areas include biomaterials, tissue engineering, industrial biotechnology, genetically engineered crops, and even biological warfare. These applied areas will likely stimulate technological convergence among innovations in physics, chemistry, computer engineering, information technologies, and other sciences.

We define biomedicine as the subset of bioscience that relates to human health, from the prevention and the diagnosis of disease to its treatment and monitoring over time. Biomedicine will spawn new treatments for debilitating diseases, vastly improved artificial limbs and organs, improved monitoring of our health, and better integrated approaches to medicine. The aim of biomedicine is to improve mankind's everyday physical and mental health, while providing affordable and highly personalized medical care around the world.

Even if the pace of scientific breakthroughs continues unabated, many factors could slow the development of biosciences. Public opinion could turn against these sometimes frightening new technologies, as they raise fundamental questions about the nature of human existence and the role of technology in our lives. Experts warn about the emergence of new diseases that could be difficult or impossible to cure, as with AIDS or the much-feared Ebola virus. Threats from bioterrorists might arise. Global power shifts, particularly the rapid growth of emerging economies, could reshape the global biomedical industry in ways that would affect the development of new treatments and cures for us all.

Although the promises are numerous and wide ranging, the ultimate impact of biomedicine on society will depend on how the underlying sciences unfold and whether successful business models can be found to commercialize them. This, in turn, will depend greatly on society's appetite for these sometimes controversial technologies (from cloning to genetic testing), including governments' willingness to fund the research and spread the benefits broadly among its citizens. Considering the scientific, technological, social, and ethical forces at play, many possible paths exist, even though only one will become reality.

Origins of this book

In this book, we try to understand which factors will especially influence the commercial potential of emerging life sciences technologies for human health between now and the year 2025. We chose this time frame because it is long enough to reveal major change and short enough to be relevant to our own lives and those of our children. We examine three overarching questions. First, how far might biomedicine take us in the future, in terms of breakthrough treatments for major public health concerns such as AIDS and malaria, cardiovascular diseases, and cancer? Second, what ethical and social questions or barriers do these new treatments and technological possibilities raise? Third, to what extent can we afford various new bioscience solutions, and how will we pay for them? The answer to each question depends on the others, so they are collectively shrouded in considerable complexity and uncertainty.

To address these fundamental questions, we launched a broad research initiative at The Wharton School of the University of Pennsylvania, called the *Future of the Biosciences*. The project started in 2002 at Wharton's Mack Center for Technological Innovation, in partnership with a management consulting firm specializing in scenario planning (Decision Strategies International, Inc). This project, which one of us directed, entailed a three-year research process of information gathering, analysis, and discussion about the critical issues facing decision makers who want to develop biomedical technologies in the coming 15 years.

This book is an outgrowth of the original Wharton project, but with a much stronger focus on *human health*. Also, it is aimed at a broader audience, extending beyond academic researchers and business leaders. We wrote this book for both intelligent laypeople and health professionals who want to stay abreast of developments in biomedicine. We present pertinent facts and explanations, along with conceptual frameworks, to help readers make sense of future developments. Compared to other books, we have sought grounding in the underlying science, as well as a broad future perspective that extends beyond the industry to society-at-large. We are still in the early dawn of the bioscience revolution, and so much more is to come.

An uncertain future

No one can credibly predict how this complex and uncertain field will unfold, just as it would have been foolhardy to attempt a precise prediction circa 1970 of where the emerging computer revolution would take us. It is humbling to recall that IBM founder Thomas Watson predicted a global demand for just four computers, and that DEC founder Ken Olsen dismissed personal computers as unnecessary in people's personal lives (who needs them?). Even such a visionary and celebrated business leader as Bill Gates didn't fully recognize the potential of the Internet until Netscape was at his doorstep, and he later failed to respond in time to threats Google posed in desktop and Internet search. Such blind spots and leapfrogging are typical in emerging technology.

Although specific predictions are bound to fail and blind spots are unavoidable, we *can* describe several broad paths that the future might take and begin to understand the key drivers involved. This can help decision makers prepare for multiple futures and build capabilities to succeed across a wide range of scenarios. Between now and 2025, the biosciences will likely become one of the most important topics in our personal lives, at work and in society. We expect major successes and failures in the commercialization of emerging life science technologies. The path ahead is fraught with risk, surprise, challenge, and opportunity. Deep trends and uncertainties will affect the outcomes, including the unpredictable actions of decision makers, the influence of public opinion and popular media, and related developments in other industries. Out of this complex concoction, various biomedical industries will either crawl or spring forth. What will they look like in 2025?

Our book offers a nontechnical introduction to the broad field of biomedicine, with an eye to the future. We first examine the current state of the field, including a short scientific history; then we explore future developments through the lens of multiple scenarios. We conclude by examining various intriguing implications of the emerging biosciences for your health, family, and career; society; and our shared global habitat. Figure 1 gives a chapter overview.

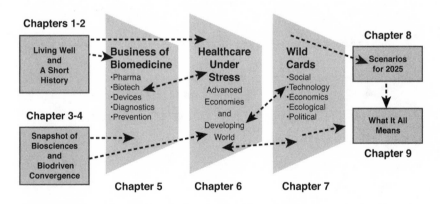

Figure 1   Schematic of chapter flows

1

Living well beyond 100

Popular culture, as reflected in movies, fashion, and literature, is pre-occupied with remaining young. A growing anti-aging industry offers myriad products and services to tap into the elusive fountain of youth, with plenty of hucksters and charlatans preying on our dreams. The quest for youth has been with us since antiquity. King Gilgamesh, who ruled parts of Mesopotamia around 3000 B.C., searched for immortality after a close friend died. To accomplish this divine goal, he had to become a half-god, according to Babylonian legend. The first historical record of a treatment to reverse aging is an Egyptian papyrus dated around 1600 B.C. that describes an ointment for regaining youth (without evidence of success or a money-back guarantee). Youth concoctions are still produced and consumed daily, but thus far, the real fountain of youth has come from science and medicine.

Great progress made, much more to come

Mankind made great strides in extending the human life span over millennia (see Figure 1.1). Life expectancy for early humans, including Neanderthals, was painfully short: Half died by the age of 20. In the nineteenth century, life expectancies rose to about 40 years in Western Europe and to 70 years by mid 20th century. These advances were largely the result of treating childhood diseases and conquering infections. Little progress had been made yet in extending the lives of the elderly. For example, American males who managed to reach the age of 70 lived only three years longer at the end of the 20th century than at the beginning of the century. In recent years, however, mankind has extended life for those who reached middle age, ushering in the age of centenarians. Today more than 30,000 people in the United States have passed the magic century milestone. Demographers expect that, in 2020, 300,000 people older than 100 will be living in the United States, with similar tenfold increases projected for centenarians in other developed nations.[1] Even though life extensions manifest themselves by people dying at an older age, this can be due to health improvements introduced at the beginning, middle, or end of life. Historically, longer life spans were achieved by focusing on the beginning and middle of life. Today, medicine is making great strides in all phases, but especially the later stages.

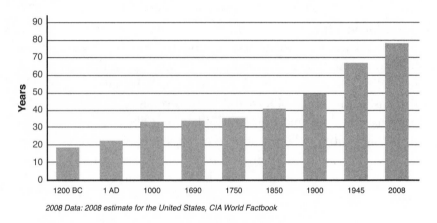

2008 Data: 2008 estimate for the United States, CIA World Factbook

Figure 1.1 Average life span in the Western world since ancient times

As shown in Figure 1-2, a 50-year-old woman living in the United States in 1990 could look forward to an average of 31 additional years of life. If we assume a cure for cancer, this increment beyond 50 grows to 34 years; adding a cure for heart disease amounts to 39 extra years. After we conquer stroke and diabetes, the increment rises to 47, yielding a full life expectancy of 97 years.[2] No one knows for sure how far we can push the boundary of death, but optimistic scientists consider a life expectancy of 130 years within reach for many young people living today.[3] The last bar shows this optimistic projection, which reflects the combined effect of many interventions, including diet and lifestyle changes.

By 2020, new approaches for diseases that presently cause death in the elderly—notably cancer, heart disease, stroke, and

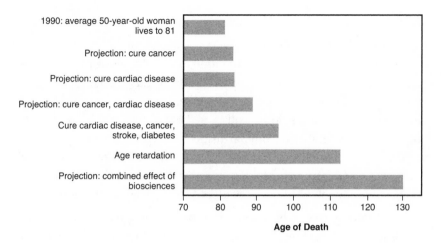

Figure 1.2   Longevity projections

diabetes—are expected to increase quality and quantity of life significantly beyond the gains made in the 20th century. Biochips offer much cause for optimism, to monitor our health more closely and eventually deliver medicines in highly customized ways. Stem cells and therapeutic cloning also show promise in regenerating damaged tissue and perhaps entire organs. In addition, new medicines might become widely available for Alzheimer's, Parkinson's, and other degenerative diseases that currently diminish the quality of life for many older people and their families. Although some skeptics in the scientific community believe that we could be reaching the limits of human longevity, other experts believe that many young people in good health today will reach the age of 100 and beyond.[4]

Extending human life expectancy

"Many experts believe that within a decade we will be adding more than a year to human life expectancy every year. At that point, with each passing year, your remaining life expectancy will move further into the future."

—*Ray Kurzweil and Terry Grossman*, Fantastic Voyage

Medical challenges and promises

The path to longevity encounters many obstacles, including the need to reset the body's built-in "regeneration clocks" that currently limit the number of times each cell can reproduce itself. Like an old car or building, parts of the body begin to wear out before we reach the end of our life spans. Skin tissue loses collagen, begins to wrinkle, and stops replacing itself. Joints become arthritic or simply wear out, requiring replacement. Plaque builds up in our brain tissue, clogging the neural pathways and contributing to degenerative diseases such as Alzheimer's that eventually rob us of our very identity and social connections.

The good news is that research into slowing aging and expanding life is in full swing. One of the first longevity genes was discovered in the tiny roundworm. In 1993, researcher Cynthia Kenyon was able to double the roundworm's life span by making a single mutation in the *daf-2* gene, just one of the 19,000 genes in this nematode's genome. Such mutated nematodes not only live longer, but also live healthier. This single mutation in *daf-2* has far-reaching effects on the worm's metabolism, its response to environmental stress, its ability to fight infection, and its control of damage by free radicals.[5] The end result of these cellular effects is a remarkable slowdown in aging.[6] Since the discovery of *daf-2*, scientists have located other longevity genes in yeast, fruit flies, mice, and rats. They have also identified human counterparts of these genes, giving hope that they can be targeted to possibly extend the life span of humans. Based on empirical results obtained from various animal models, researchers hope to extend human life by 30% or 40%, while at the same time reducing the disabilities of old age.

Caloric restriction, also described at times as a starvation diet, is another active area of research. It examines how reducing the consumption of calories (while still eating nutritionally balanced meals) slows aging and increases life span in mice, rats, dogs, and primates. Animals who start such a diet at a young age increase their life span 30% to 50%. Calorically restricted animals exhibit much higher activity levels, experience fewer common diseases such as cancer and heart disease, and appear more youthful and energetic overall. They spend the last part of their lives in relatively good heath, without the major infirmities of aging. One major downside in some animals, however, is a loss of fertility (as happens with very underweight female athletes). Scientists have known about the benefits of calorie restriction for nearly 70 years, but how it works has only come to light in the last decade. In yeast and worms, the *SIR2* gene seems to influence longevity. Researchers have identified a similar gene in flies, mice, and humans.

Few humans would voluntarily sign up for a 1,000-calorie-per-day diet for the unproven promise of living considerably longer. So, pharmaceutical companies such as Sirtris, Novartis, and Elixir are designing compounds that mimic the effects of mutations in the *SIR2* gene. These companies look for drugs that affect the activity of the Sir2 protein and related proteins called Sirtuins. One such compound, resveratrol, exists in red wine. Feeding resveratrol to yeast, worms, or flies extends their life span about 30%.[7] The hunt is on for even more potent compounds that could slow the ravages of aging further while at the same time treating diseases associated with aging, such as Type 2 diabetes, cancer, and cardiovascular decline.

We expect to see much more research aimed at slowing, or perhaps even reversing, the aging process. These investigations run from basic molecular bench science about the biology of aging, to epidemiological research on the traits and life styles of centenarians, to animal studies of the kind mentioned earlier. Researchers continue to explore important avenues and report promising advances that could extend human life to 120 or beyond:

- **Replacements for aging cells**—The ability of embryonic stem cells to morph into almost any type of cell, from heart to liver or brain cells, holds the promise that they can replace damaged, diseased, or destroyed cells. This includes cells lost as part of the normal aging process, as well cells damaged by stroke or heart attacks. Scientists have cloned many therapeutically important proteins in the laboratory, including insulin and human growth hormones, to replace our own hormones. Eventually, this will lead to tissue regeneration and perhaps the replacement of entire organs.

- **Improved diagnostics**—The sequencing of the human genome and the continual improvements in this technology have made it possible to find the single nucleotide polymorphisms (SNPs) that represent alterations or mutations in our genes. These distinctive gene alterations are important because they could affect our susceptibility to disease. By 2025, millions of SNPs could be linked to some of the most vexing diseases, such as cancer, heart disease, Alzheimer's, and Parkinson's. Diagnostics based on this wealth of genome information could become common. (See Appendix C, "Complexity of the Genome," for more information.)

- **Targeted preventive therapies**—Better genomic information could lead to more effective therapeutic and lifestyle interventions, such as changes in diet prescribed to prevent the onset of disease. In more complicated diseases, bioinformatics is already paying dividends by assessing the role of genetics and both environmental and lifestyle factors. Owning your own genetic information on a microchip will become so affordable in the future that many people can practice preventative medicine by starting individualized therapies at earlier ages, as suggested by their genes.

- **Snap-in knee joints**—In Paoli, Pennsylvania, arthroscopic surgeon Dr. Kevin Mansmann has been working on materials and procedures to enable arthroscopic insertion of replaceable pads to cover worn surfaces in bone joints, in the knee and other joints. Even if they wear out, these synthetic components could be easily replaced with a surgical procedure that could become routine.

- **Fantastic nano-voyage**—Novelists and screenwriters have mused about taking a robotic 'fantastic voyage" through the human bloodstream. Nano devices, which operate at a scale of 1/1 billion meters, likely will be inserted into our bodies as sensors and biochips. One day, nano-robots may have the capability to detect, clean, or repair damaged blood vessels or organs from the inside, like a mini medical roto-rooter.

- **Altering genes**—Researchers are using molecular medicine to unlock the secrets of the aging process itself, by trying to uncover the genes involved. In the field of plant genetics, developers of the first genetically engineered plant (the Calgene tomato) were able to keep the tomato fresh longer by shutting down the gene that causes the tomato to rot and release its seeds. If scientists can do this with tomatoes, perhaps they can manipulate aging genes in humans, whether rotten or not.

In an era in which body parts can be replaced as they wear out, we need a new way to think about aging itself. Will there come a day when someone asks your age and you respond by saying, "I was born 91 years ago, but my knees are 16, my heart is 20, and my kidneys are 14 years old"? Like an airplane in which most parts have been replaced, with only the basic main frame still being original (but repaired and reinforced in places), how do we assess its real age?

Living well versus living longer

The relevant issue is not only how long we live, but how well we live. In medical terms, this means looking at morbidity (which refers to illness, disability, and disease) as well as mortality. Much will depend on the way a longer life span is achieved, which can happen through compressed morbidity, decelerated aging, or arrested aging. In the first case, people live well into old age, relatively free of disease, and then die within a short period of time. The second profile represents a general slowing of aging, with an extension of the old age period (as is attempted today in developed nations). The third possibility represents a stopping or even reversing of the aging process, such as by resetting the biological clock, which could result in restored vitality and rejuvenation among the old (as in the movie *Cocoon*). Figure 1.3 depicts these three profiles of morbidity, recognizing that many other shapes are possible in years to come. For completeness, we add a fourth profile, which is still prevalent in developing nations and historically represents much of the human experience. This profile depicts a rather short life burdened by disease and illness for a major portion of the person's time on Earth, assuming they are lucky enough to reach maturity in the first place.

Many of the improvements in morbidity are due to reductions in death from childbirth or childhood illness, as well improvements in sanitation (see Chapter 2, "A Short History of Biomedicine"). One of the most valuable medical inventions is the lowly toilet, which helped improve hygiene and sanitation.[8] Over the past 200 years, however, an additional factor has been the overall improvement in the size, health, and robustness of human beings. In earlier times, malnutrition affected the unborn, children, and adults alike, making them all susceptible to infections and maladies that stronger humans can bear. Path-breaking research by Robert Fogel, an economic historian who

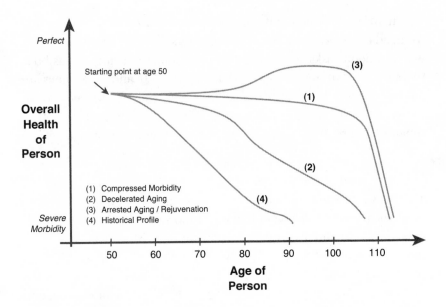

Figure 1.3 Quality of life profiles

won the Nobel Prize in economics in 1993, underscores the importance of what he termed "technophysio" evolution. Fogel's team painstakingly studied the medical records of about 45,000 Union Army veterans and then compared this Civil War generation with soldiers in World Wars I and II. Using detailed medical records, pension files, and death certificates, they examined morbidity and mortality over a century.[9] The research shows how humans were gradually liberated from centuries of malnutrition, resulting in a doubling of life expectancy since 1700 and an increase in body mass of more than 50%. For example, an average Frenchman alive in 1790 weighed around 110 pounds when in his 30s, compared to weighing more than 170 pounds today. Since 1775, Norwegian men have added 5.5 inches to their height. In India, life expectancy at birth has risen from 29 to 60 since 1930.

Of children born around 1840, about 25% died in infancy and another 15% died before they turned 15. Those who reached adulthood suffered from persistent malnutrition. This contributed to chronic conditions such as back pain, diarrhea, and cardiovascular disease, as well as acute infections such as typhoid, tuberculosis, measles, rheumatic fever, and malaria (which was endemic throughout the southern United States). In addition, dental problems, including loss of teeth, were common. These various health stresses took their toll, especially because most jobs entailed tough physical labor. According to Fogel's records, Union Army veterans suffered an average of 6.2 chronic conditions in their mid- to late 60s (if they lived that long), compared to fewer than two such conditions today for the same age group. Over the past 200 years, humans have become physically stronger, thanks to better nutrition, fewer infections, better sanitation, and safer work environments, resulting in a species better able to withstand the medical slings and arrows of life.

Apart from objective measures of morbidity, how we feel and experience the quality of our life is ultimately most important. Quality of life perceptions are hard to measure because they necessarily rely on subjective reports by respondents. People who end up in a wheelchair, for example, might report being happier after a while than they were before. Elderly people often consider the golden years their finest, even though they are burdened with illness, disability, and the deaths of loved ones. Social scientists measure perceived quality of life in several ways, including survey questions and interviews. They validate their measures by correlating them with, say, depression or suicide statistics, to arrive at yardsticks that can be compared over time and across cultures. An important proxy that researchers use is *Subjective Sense of Well-being* (SEW), a prominent component of people's perceived quality of life. Empirical studies have identified many variables that impact people's SEW.[10] An important caveat is that many of these variables interact with each other and can vary considerably across subpopulations, time periods, or countries.

Demographic characteristics such as age, gender, and race all matter in people's sense of well-being. SEW tends to peak around age 65. Women report higher SEW scores than men before age 45, but lower scores than men later in life. Race seems to be a minor effect after adjusting for factors that correlate with race, such as education, income, and expectations. Education, income, and marital status are all positive influences on SEW. In terms of health as well as SEW, it pays to be married or in a loving relationship of mutual support. Not surprisingly, health is the most important influence in SEW, especially among older people, because the young often take their health for granted. Physical and social activities are also important, with special weight given to social integration through family, work, religion, and volunteerism. Social ties and group support matter especially during illness and hardship. A sense that one has control over one's life contributes strongly to SEW, as does the belief that one's life has meaning (for whatever reason).

Our sense of well-being is usually measured relative to others', and so much depends on who we compare ourselves to. Upward comparison might make us feel inadequate but can be a source of inspiration, motivation, and energy. Downward comparisons create a sense of satisfaction but might dull our ambition and desires. Early in life, upward comparison can help people realize their potential more fully; later in life, we need to accept who we are and what we have achieved, so a convergence between expectation and reality becomes important for psychological well-being. Lastly, SEW is influenced by the various life resources we accumulate over time, such as financial means, social support, and professional networks, as well as a sense of peace and wisdom that often comes only with age. Although it is beyond our scope to assess how the biosciences impact our perceived quality of life (for example, through better psychotropic drugs), we must recognize that health is just one factor, albeit an important one, in helping us live well beyond 100.

Social challenges and promises

Assuming that science and medicine deliver the benefits that many hope for, society will face a host of challenges for which it is not well prepared. These range from economic to moral and cultural problems, as we briefly discuss in this section. As longevity becomes more real, policymakers and public opinion leaders will increasingly force a debate about the moral and ethical issues posed by an inverted age pyramid. Bioethicists and other opinion leaders have raised several cautions about investing massive societal resources in life extension.[11] First, the notion of a finite, staged life cycle is essential to being and feeling human. Just as an arrested childhood can hamper social development, disturbing the natural rhythm of life or making it infinite might be dehumanizing. Aging is not a disease, they argue, but a normal human development. A second argument made is that family relations would be perturbed and complicated if five or more generations were alive at one time. In many ways, age is relational, defined by our roles as child, adolescent, young adult, sibling, parent, and grandparent. Extending life could upset the many social institutions that are currently tied to age, such as childbearing, schooling, military or civic duty, marriage, retirement, prison sentences, career paths, or savings habits.

A third argument concerns the social injustices caused by an older population, in terms of the disproportionate health and retirement resources consumed by these economically less productive individuals. A large senior cohort would have increased political clout, thanks to their voting powers and higher wealth levels, resulting in a biased agenda for healthcare, retirement facilities, and pensions. In addition, uneven access to life extension opportunities across the socioeconomic spectrum would likely raise moral concerns, especially if a small privileged elite emerges for whom longevity is an exclusive luxury good. A fourth argument is that the quest for longevity will lead to genetic manipulations that could change our human essence, including the slippery slope of eugenics. We stand at the threshold of an unprecedented era in which humans can change their own genes, and hence redefine what it means to be human. Unfortunately, we presently lack the regulatory oversight and moral compass to wisely navigate the technological terrain.[12]

Can we afford old age?

Assuming that science indeed offers us the possibility of significant life extension, can society afford it? We list here some of the key challenges currently facing our aging society in the United States.[13] By understanding the major challenges facing the richest nation on Earth today, we can start to appreciate the numerous economic, social, and moral burdens posed by a much older society, especially one in which many people live beyond 100.

- **Long-term care financing deficit**—The current U.S. system of paying for long-term care is mired in inconsistent regulations and legislation that fragment health and social services for the elderly. Medicaid is the largest public payer of long-term care services, but individuals qualify only if they are impoverished. Medicare is the second-largest public payer, restricted to those over age 65. It will explode the U.S. federal budget in years to come. The largest source of private financing is out-of-pocket payments by those who need care. The next source is private insurance, which finances less than 7% of long-term care.

- **Insufficient resources for assisted living**—Assisted living is the fastest-growing segment of the aging services continuum. More than one million people reside in assisted-living environments in the United States. Medicaid currently covers the cost for only 2% of these residents, with the rest paid via insurance, through other programs, or privately. Monthly rates ranging from $2,000 to $4,000 make this type of care unaffordable for the poor and near-poor. Older adults often want to remain in their own homes as long as possible. Funding for home services is fragmented and often inadequate to meet demand.

- **Staffing crisis**—Service providers for the elderly are experiencing an acute staffing shortage. Nursing homes, assisted-living facilities, and home-health agencies all require qualified staff to provide suitable-quality care and services. Yet they are hampered by a tight labor market, noncompetitive wages, and benefit levels limited by current Medicare and Medicaid reimbursement rates. Negative public perceptions about the field and limited opportunities for career advancement further exacerbate the problem. Accusations of abuse and neglect of the elderly by caregivers are increasing. In addition, geriatrics fellowships for training doctors remain open, while more lucrative fields such as dermatology attract young doctors.

- **Liability insurance crisis**—State laws provide nursing home residents with a bill of rights, and rightly so. But this has also spawned the growth of law firms specializing in nursing home liability claims, with the danger of excessive litigation, as has happened in medicine generally in the United States. Elder law will become a growing field to protect senior citizens and could impose another large litigation burden on society if unchecked.

- **Informal and family care**—The typical caregiver for the elderly is a married woman in her mid-40s who works full time, providing care to one person, most likely a relative. These informal caregivers often provide intense care that involves extensive personal and financial sacrifice, as well as physical and emotional stress. The impact on the workplace is enormous, with absenteeism among caregivers on the rise. Corporate America loses more than $11 billion a year because of absenteeism, turnover, and lost productivity among full-time employees who care for elderly people.

Government and business

As we contemplate the promises of different biomedical technologies and the social challenges they create, we must consider the crucial role of business and government in either supporting or retarding progress. Governments set the rules of the game, in terms of reimbursement, taxation, research incentives, intellectual property rights such as patents, antitrust legislation, import/export restrictions, and fiscal and monetary policy. When the Bush Administration imposed a moratorium on embryonic stem cell research in 2001, the stock-market value of many stem cell companies dropped sharply and scientific talent started to explore research opportunities outside the United States, including those in Europe, South Korea, and Singapore.[14] Conversely, government can stimulate applied research, as happened with the celebrated Bayh–Dole Act of 1980.[15] This seminal legislation, plus key amendments in 1984 and 1986, allowed universities, other nonprofits, and small companies to own the commercial rights to their research even if the federal government funded it via basic research grants. This, in turn, allowed NSF- or NIH-funded research to be licensed to industry, giving biotech and pharmaceutical companies access to promising technologies that would otherwise remain dormant in academia. Another positive example of government support for the biosciences is the multibillion-dollar Biopolis project funded by Singapore (see the accompanying sidebar). This strategic initiative epitomizes an important partnership between government and business, and illustrates that the field of the biosciences has a wide global footprint, from basic research to the eventual delivery of clinical benefits to patients.

Whereas governments can set the rules and provide incentives, the free market is especially vital in bringing new medical technologies to patients. Scientific discoveries and medical inventions can languish for decades if no viable business model commercializes them.

Singapore: the biopolis of Asia

The government of Singapore helped the pharmaceutical and biotech industry by setting up the Biomedical Sciences Initiative in 2000. Since then, the sector has taken off. Its manufacturing output rose to about $20 billion in 2008, an almost fourfold increase from 2000. Singapore's ability to quickly put together the various resources and infrastructure needed to attract foreign investors and grow the industry has proved key. The crown of this achievement is Biopolis, a still-evolving and growing R&D complex that houses the country's major public biomedical sciences research institutes, as well as private-sector R&D laboratories.

The co-location of private companies and public research laboratories at the Biopolis is deliberate. This arrangement enables the entities to share research facilities, equipment, and amenities, which helps to overcome a major challenge that both start-ups and established companies face: the need to manage R&D costs and shorten time-to-market. At Biopolis, tenants can take advantage of its "plug-and-play" infrastructure and access shared facilities and state-of-the-art equipment, including X-ray crystallography, nuclear magnetic resonance, electron microscopy, DNA sequencing, and more. Beyond the physical infrastructure, Biopolis has attracted an impressive number of top scientists who are based there. Singapore has adopted a "queen bee" approach to building its research base by enticing key international players and assumes that others will follow.

—*Contributed by Marvin Ng, DN Venture Partners, Singapore*

The key distinction between invention and innovation is this commercial link: Innovation is all about finding a way to convert inventions into products and services that can be profitably brought to market. Figure 1.4 highlights two key inflection points in the evolution of technological innovations.[16] The first concerns a scientific battle among competing modalities in the early stages of a new technology. A classic example is the struggle between alternating current (AC) and direct current (DC) in the days of Thomas Edison, which lasted several decades and was finally settled in favor of AC power. After the scientific questions are settled, through research, testing, and peer reviews, a second inflection point usually occurs around competing models for application in the market. A well-known example is the fierce battle in the 1970s between Sony's Betamax and JVC's VHS format for the home video and entertainment market, or later between Apple and Microsoft in setting the common standard for operating systems in personal computers. A current biomedical example concerns the competing standards being offered for electronic medical records. Sometimes these battles amount to a winner-take-all contest, especially when there is room for only one main standard (such as AC versus DC power) or when special protections are granted, via patents or preemptive contracts, to the first mover (as with blockbuster medical drugs).

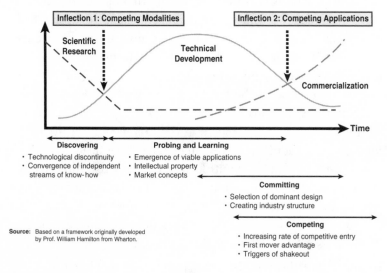

Figure 1.4 How technologies evolve

The importance of business, especially the crucial role of business models, is central in understanding the evolution of the practical applications of the biosciences. By *business model*, we mean the broader conceptual framework that companies use to create, produce, and deliver products or services to the marketplace. For example, a traditional bookstore chain such as Borders in the United

States operates under a different business model than Amazon, which relies heavily on the Internet. Not only are their internal operations—from procurement to warehousing, to shipping—different, but so is the way they market and distribute their books and other products. Akin to a species that has adapted to a specific ecology, the business model represents the strategic blueprint by which a company or entire industry manages to be viable, given its external environment.

To become viable, new technologies that do not fit existing business models either will remain dormant or must find new business models, via entrepreneurs or innovations by existing players. If such new technologies destroy previous business models, as supermarkets have done to local grocery stores and the Internet has done to the traditional travel agency, business scholars term them *disruptive technologies*.[17] The many technological changes afoot in the biosciences likely will prove disruptive to existing pharmaceutical, diagnostics, and medical device companies. Just as the personal computer revolution led to the demise of the traditional vertically integrated mainframe computer companies (such as IBM, DEC, Wang, Nixdorf, and NEC), the personalized medicine revolution could result in the breakup of the traditional integrated pharmaceutical companies (such as Pfizer, Merck, GlaxoSmithKline, and Novartis). Although the industry is presently consolidating, large companies eventually might restructure around areas of genuine core competence and spin out the activities that no longer bestow competitive advantage. The rise of new business models will force this restructuring.

To appreciate the promises and pitfalls of the biosciences, we must understand the complex dynamics among technological progress, social needs and concerns, and governmental oversight and regulation. As depicted in Figure 1.5, these three areas can be viewed as the corners of a complex triangle, with back-and-forth interactions among the corners as they change over time. In the middle of this triangle lies the continually evolving world of business models that try to connect the corners to produce valuable products and services for consumers. This is where both existing companies and entrepreneurs look for opportunities. Ultimately, however, major innovations in biomedicine hinge on two deeper questions: How far can the underlying sciences really take us, and do we want to go there as a society?

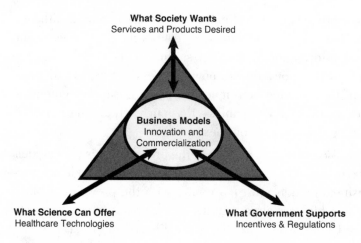

Figure 1.5 Business models tie it all together

The journey ahead

Although many of us desire better and longer lives, the collective burden of this wish could become a curse on society. We explore in this book the scientific and technological forces that make living beyond 100 possible, as well as the economic, social, and political obstacles that might stand in the way. The interplay of these forces and obstacles will shape the world between now and the year 2025. As educated consumers or professionals connected to the biosciences in its many facets, we need to understand this dynamic in greater detail. It will prepare us to participate in increasingly important public debates, as well as guide us in making wise health choices for ourselves and our families. In addition, those involved in organizations that create, deliver, or support healthcare services need to be aware of important undercurrents that can reshape markets and even entire industries in short order, akin to what we saw in information technologies and the computer industry over the past four decades and more recently in banking.

The next chapter offers a short history of biomedicine covering the past several decades; future developments in the biosciences rest on the shoulders of this remarkable era. Chapter 3, "Snapshot of the Biosciences," provides an overview of bioscience technologies and delineates more clearly our limited focus on human healthcare. Chapter 4, "Bio-Driven Convergence" reviews the confluence of life science and information technologies; this intersection holds the promise of great innovation via new business models. Chapter 5, "Business of Biomedicine," examines the major business segments—from pharma to device companies—that currently define biomedicine, including their current business models.

New biodiscoveries will need to fit the prevailing business models or find alternate routes to the market and patients. In addition, successful business models must harmonize with the prevailing healthcare systems, which are already under much pressure, as Chapter 6, "Healthcare Under Stress," examines. Healthcare, in turn, is part of a much larger socioeconomic and political system, in which numerous macro forces comingle to shape our future global environment. We describe these larger forces in Chapter 7, "Wildcards for the Future," with a special focus on uncertainties that can really rock biomedicine. Chapter 8, "Scenarios up to 2025," presents different future scenarios that could result from the interplay of these macro forces. These scenarios reflect the inherently unstable balance between numerous technology drivers on one hand and various economic, social, and political forces on the other hand. Lastly, Chapter 9, "What It All Means," summarizes what all the push-and-pull means for you personally, your family, your work life, the field of commerce more broadly, and society at large.

Endnotes

[1] According to the U.S. Census Bureau, the number of centenarians has grown 60% since 1990 and will continue to rise to 274,000 by 2025.

[2] Richard A. Miller, "Extending Life," chap. 10 in *The Fountain of Youth*, Stephen. G. Post and Robert. H. Binstock (eds), (New York: Oxford University Press, 2004).

[3] Ray Kurzweil and Terry Grossman, *Fantastic Voyage: Live Long Enough to Live Forever* (Emmaus, PA: Rodale, 2004); In *Ending Aging: The Rejuvenation Breakthrough That Could Reverse Human Aging in Our Lifetime* (New York: St. Martin's Press 2007), Aubrey de Grey argues that new technologies will extend our lives significantly.

[4] In *More Than Human: Embracing the Promise of Biological Enhancement* (New York: Random House, 2005), Ramez Naam explains why science is on the verge of extending life, including creating designer babies, and that we should embrace this progress ethically as well as otherwise; see also William Hanson, *The Edge of Medicine* (New York: Palgrave MacMillan, 2008).

[5]Free radicals are chemical compounds inside cells that are electrically charged. They are routine byproducts of metabolism, mostly from burning fuel using oxygen. In their hunt for electrons, free radicals can mutate DNA when they strip electrons from it, leading to cancer and other diseases.

[6]Stuart K. Kim, "Proteins That Promote Long Life," *Science* 317 (3 August 2007): 603–604.

[7]David A. Sinclair and Lenny Guarente, "Unlocking the Secrets of Longevity Genes," *Scientific American* (March 2006): 48–57.

[8]Rose George, *The Big Necessity: The Unmentionable World of Human Waste and Why It Matters* (New York: Metropolitan Books, 2008).

[9]Robert Fogel, *The Escape from Hunger and Premature Death, 1700–2100: Europe, America, and the Third World* (New York: Cambridge University Press, 2004).

[10]Linda K. George, "Perceived Quality of Life," chap. 18 in *Handbook of Aging and the Social Sciences* (St Louis, MO: Academic Press, 2006).

[11]The following are well-known critics of prolongevity: Daniel Callahan, Francis Fukuyama, and Leon Kass. See Robert Binstock et al., "Anti-aging Medicine and Science," chap. 24 in *Handbook of Aging and the Social Sciences* (St Louis, MO: Academic Press, 2006).

[12]Francis Fukyama, *Our Post-human Future: Consequences of the Biotechnology Revolution* (New York: Farrar, Strauss & Giroux, 2002).

[13]This study was conducted through the consulting firm Decision Strategies International, Inc. (www.thinkdsi.com), and resulted in two scenario reports about the future of aging. Both reports are available from the American Association of Homes and Services for the Aging (AAHSA) in Washington, D.C. (www.aahsa.org).

[14]Specifically, President George W. Bush announced on August 9, 2001, that federal funds may be awarded for research using human embryonic stem cells if three criteria are met: (1) The stem cell lines were already in existence, (2) the stem cells were not derived from an embryo created for reproduction, and (3) informed donor consent was obtained.

[15]This legislation was co-sponsored by Senators Birch Bayh of Indiana and Robert Dole of Kansas, and enacted in 1980. The Bayh–Dole Act created a uniform patent policy among U.S. federal agencies that fund research, allowing small businesses and nonprofit organizations, including universities, to retain own, trade, and license inventions made under federally funded research programs.

[16]This section draws on research conducted at our Mack Center for Technological Innovation at the Wharton School, University of Pennsylvania. For details, see George S. Day and Paul J. H. Schoemaker, eds., *Wharton on Managing Emerging Technologies* (New York: John Wiley & Sons, 2000).

[17]Richard Foster, *Creative Destruction* (New York: Doubleday, 2001); Clay Christensen, *The Innovator's Solution* (Boston: Harvard Business Press, 2003).

2

A short history of biomedicine

We've come a long way in improving human health. The last five decades have revealed more about the biology of life than any other period before. This chapter sketches remarkable historical triumphs in molecular biology and related sciences. We need this understanding to appreciate the foundations on which biomedicine rests today and builds tomorrow.

Before the twentieth century, life was brutal, disease ridden, and often short. The major causes of death were childbirth and all kinds of infectious diseases, the latter often coming in the form of plagues that killed millions (see the following sidebar).

Malnutrition was especially rampant and greatly increased people's susceptibility to disease. In the 1800s, people thought bad air caused deadly diseases such as cholera and the Black Death. No one knew yet what roles microbes played or how illness could spread. Ironically, hospitals were often a breeding ground for disease instead of a safe haven, due to limited understanding of how infections were transmitted.

Improving hygiene

In the early 1800s, death during childbirth—called child bed fever or puerperal sepsis—was quite common. Ignaz Semmelweis, a young doctor who ran the maternity ward of a hospital in Vienna, wondered why the death rate of mothers was as high as 20%—four times higher than in the ward where midwives looked after expectant mothers. Later he noticed that a fellow doctor had died from puerperal sepsis after cutting himself during an autopsy of a victim of child bed fever.

Killer epidemics

Epidemics became more common as humans migrated to new destinations, either carrying diseases that impacted natives or becoming victims themselves due to a lack of immunity. The Black Death, a pandemic that raged especially from 1347 to 1351, devastated Asia, Africa, and Europe. It killed about 100 million people worldwide, with Europe losing one-third of its population and China up to half. This makes the Black Death the worst known nonviral pandemic in history. Not until 1894 did Alexander Yersin and Shibasaburo Kitasato independently discover that a bacterium, later named *Yersinia pestis* in honor of Yersin, was the culprit.[1]

Slowly, the epidemiology of the disease was unraveled: how biting flies carried it from infected rats to humans, aided by the dirty living conditions of the day. Epidemics continued unabated into the twentieth century, with smallpox, tuberculosis, cholera, and typhoid exacting a heavy toll on human populations throughout the world, culminating with the influenza pandemic of 1917–1919. Twice as many people died in this pandemic than were killed in World War I, which was also raging at the time. The death toll of this influenza was between 50 million and 100 million worldwide, with about a half-million victims in the United States alone.[2]

This led him to speculate that a disease-causing agent might inadvertently be carried from the autopsy rooms to the birthing rooms, infecting mothers and causing the higher rates of child bed fever. Based on this conjecture, he insisted that doctors wash their hands and change their contaminated clothing after autopsies. Death rates dramatically dropped. Unfortunately, his fellow physicians resented the new rules and the implicit suggestion that they were spreading the disease. Dr. Semmelweis was soon fired. His innovations in hygiene were never fully appreciated during his lifetime, and he died a dejected man in an insane asylum. Gradually, his hygienic principles became standard practice in hospitals, together with those of English surgeon Joseph Lister, who pioneered the use of antiseptic agents for wounds and heat sterilization of surgical instruments.[3]

Another remarkable scientist of the period, English physician Dr. John Snow, tied the incidence of cholera during the 1854 London epidemic to contaminated water. Londoners received their water from two different companies: one upstream and one downstream of the city. By tracing the disease progression, Snow was able to show that people served by the downstream company, whose water was contaminated with human sewage, experienced cholera far more often than those served by the upstream one. This led Snow to recognize that cholera was spread by water. He also associated a localized epidemic of cholera with a water pump serving the Broad Street area of London. Snow meticulously plotted the history of the cholera cases and showed how they centered on, and radiated concentrically from, the Broad Street pump. Removing the pump's handle brought an end to the epidemic.[4] Snow's discovery was a breakthrough in public health, and he is considered one of the fathers of epidemiology.

In 1900, infectious respiratory diseases such as pneumonia, influenza, and tuberculosis were the leading causes of death in the United States. But by 1910, heart disease had risen to take that dubious honor.[5] It remains the leading cause of death in the richest countries, followed by cancer, although respiratory infections still account for most deaths in the poorest countries.[6] Infectious diseases experienced a steady decline in the developed world over the last century due to some major scientific breakthroughs that we address next.

The power of immunology

In the 1790s, English physician Edward Jenner was the first to discover a vaccine for smallpox, one of the most dreaded diseases before 1900. Smallpox killed about one-third of those who became infected. Jenner noticed that milkmaids who got cowpox, a non-contagious disease that causes nasty sores or pustules on the hands from touching cows, never became infected with smallpox. He conjectured that the exposure to cowpox somehow bestowed resistance to smallpox infection. He tested this theory by deliberately infecting a young boy with material taken from the cowpox sore of a milkmaid. A few months later, he exposed him to deadly smallpox material. Infecting young boys would surely not pass muster today, but these were different times with different ethical norms. Luckily, the young test patient survived this exposure and 20 more such exposures over the following 25 years, confirming Jenner's theory. Jenner's spectacular results led to widespread vaccination, with even President Thomas Jefferson and his family being vaccinated.[7]

French scientist Louis Pasteur carried the field of immunology even further. He was studying cholera by injecting chickens with the bacteria. When using a fresh culture of bacteria, most chickens contracted the disease; when using an old culture the chickens did not get sick. Pasteur assumed that old bacterial cultures had lost their ability to cause disease. But when he took chickens that were previously infected with an old culture, and then infected them with a fresh new batch, they failed to show any symptoms of cholera. This puzzling result led Pasteur to realize that even though the old bacterial culture had become attenuated, or weakened, it could still protect against new infections with virulent cholera by activating the immune system. Since this key discovery in the 1870s, this attenuation process has become the basis for how many vaccines are produced.

Monoclonal antibodies

In 1975, English scientists Cesar Milstein and Georges Kohler devised a laboratory method for obtaining large quantities of highly specific antibodies originating from a clone of B lymphocytes. These antibodies, termed *monoclonal antibodies*, are highly targeted protein molecules that attack foreign antigens on bacterial or viral invaders and tumor cells. However, B lymphocytes do not survive very long in culture. To solve this, Milstein and Kohler fused these antibody-producing lymphocytes with tumor cells that are able to divide indefinitely in culture. The resulting cell, called a *hybridoma*, combines the traits of each cell and can churn out antibodies of a single specificity continuously in culture (see Figure 2.1).

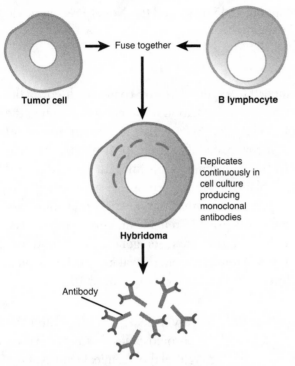

Figure 2.1   Making monoclonal antibodies

The discovery of antibiotics

The landmark discovery of penicillin by British bacteriologist Alexander Fleming in 1928 is one of the most notable examples of the power of a prepared mind in science. Chance favored this keen observer when he noticed that a common mold that had accidentally fallen onto one of his bacterial cultures killed the bacteria growing nearby. Fleming went on to make an extract from the mold and showed that it protected mice from bacterial infection without killing them. Thirteen years later, in 1941, Howard Florey and Ernst Chain developed a form of penicillin that could be used in humans.

Penicillin became the most widely used antibiotic in history and initiated the era of modern antibiotics. This wonder drug saved countless wounded soldiers from death in World War II and many millions more thereafter. It proved especially valuable in killing staphylococci which infects wounds, and other pathogens such as those causing scarlet fever, pneumonia, gonorrhea, meningitis, and diphtheria. Fleming was knighted in 1944 and in 1945 he shared the Nobel Prize in medicine with Florey and Chain.

The DNA revolution

Many scientists pursued the secret of life by trying to discover the molecule(s) that carry genetic information from one generation to the next. The search started in the 1860s with the elegant studies of Augustinian monk Gregor Mendel. Using pea plants, Mendel painstakingly recorded the inheritance of distinct traits such as seed color, seed shape, and plant height over multiple generations. He realized from crossing pea plants with different traits that some factors passed from parent to offspring in reliable and predictable ways. Mendel had no idea that these factors—or genes, as we now call them—consisted of DNA. Mendel's groundbreaking work was published in 1866 but attracted little attention until almost 16 years after his death.

In the last half of the 19th century, researchers elucidated the structure of the cell nucleus and the chromosomes that reside within it. In 1871, Friedrich Miescher highlighted the chemical nature of the nucleus. He isolated a large acidic, phosphate-containing material that he called nuclein, for its location. Nuclein was a major component of chromosomes that scientists intensely studied under the microscope. Biologists noted how chromosomes were passed from one generation to the next in a similar manner as Mendel's traits, and some speculated that nuclein might therefore be the heredity material.[8]

In the early 1900s, DNA was shown to consist of building blocks called *nucleotides*.[9] At the time, DNA was not considered the likely molecule of inheritance. Protein, the other known component of chromosomes, was the favored molecule. Proteins consist of amino acids; at least 20 had been discovered, whereas DNA had only 4 different nucleotides. In view of its more complex structure, protein was thought to be more capable of storing the vast amounts of information needed for millions of inherited traits than the simpler DNA structure with just four different nucleotides. However, the protein model of inheritance fell into doubt in the 1940s due to some crucial experiments, as described further in Appendix A, "DNA, RNA, and Protein."

In all, it was an arduous journey to understand how traits are passed from one generation to another, culminating in the discovery of DNA as central to this process. Figure 2.2 shows the basic insight that the DNA double helix unwinds in order to replicate and pass on genetic traits. During this process, nucleotides are assembled to form new daughter DNA molecules. Appendix A further describes how DNA's structure was elucidated, how the genetic code was "cracked," how gene expression is regulated, and how proteins are made. The discovery of DNA as the cell's storehouse of genetic information ushered in the modern era of molecular biology, unleashing a flurry of breakthrough developments in the biosciences and human healthcare.

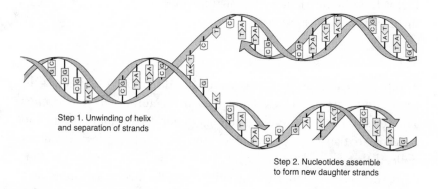

Step 1. Unwinding of helix
and separation of strands

Step 2. Nucleotides assemble
to form new daughter strands

Source: http://en.wikipedia.org/wiki/File:DNA_replication_split.svg

Figure 2.2   DNA replication

Recombinant DNA and cloning

Research in various laboratories during the 1960s yielded breakthrough techniques that laid the groundwork for the biotech revolution. DNA in chromosomes is unwieldy in terms of size. The largest human chromosome has about 250 million nucleotides, and the smallest contains around 40 million nucleotides. Even the genome of the lowly *E. coli* bacterium, the workhorse of molecular biology research, presents a significant challenge. It has only one chromosome: a continuous loop of 4.6 million nucleotides that contains about 4,000 genes. To work with DNA, scientists needed to find ways to break up the genome into fragments of more manageable size.

Enzymes that cut DNA

Enzymes would become crucial tools in investigating and manipulating chromosomes, both for scientific study and to produce life-saving biomedical treatments. All living creatures utilize enzymes to manipulate their own genetic material. Therefore, a source of these precision tools was literally right under the noses of scientists in the cells they were studying. One by one, these enzymes were isolated from cells, starting with enzymes for replicating DNA, such as DNA polymerase and DNA ligase, followed by those that cut DNA at precise nucleotide sequences. The latter, so-called *restriction enzymes*, were discovered in the 1960s in bacterial cells infected by bacterial viruses, where scientists observed them cutting up and destroying the viral DNA, thereby restricting its growth. Appendix B, "Cloning Genes," further explains this process.

Self-replicating plasmids

Another breakthrough came with the discovery of small, self-replicating, circular DNA molecules in bacteria, called *plasmids*. They exist separately from the cellular genome and can replicate and transfer to daughter cells every time the cell divides. Scientists noticed that some bacterial cells were naturally resistant to antibiotics, such as penicillin. In some cells, the resistance was conferred by a plasmid found in the cytoplasm. These plasmids were much smaller than the chromosome and often contained only a few genes, including ones that coded for enzymes that destroy antibiotics.

In 1971, Stanley Cohen of Stanford University developed a method to coax bacteria to absorb small plasmids containing antibiotic-resistance genes through their cell walls. When they absorbed the plasmid DNA, a process called *transformation*, the bacterial cells became resistant to the antibiotics due to the antibiotic-destroying proteins encoded on the plasmids. In fact, the bacteria contained multiple copies of the plasmids; the more copies, the greater the resistance to antibiotics. Therefore, plasmids provided a means to amplify the amount of a gene and its protein product. Making identical copies of a gene this way is termed *cloning*.[10]

Enzymes to cut (restriction enzymes) and paste (ligase) DNA, together with plasmids that carry antibiotic resistance, are major components of *recombinant DNA technology*. This toolkit enables *genetic engineers* to recombine stretches of DNA from the same or different organisms into a hybrid, *recombinant* molecule. Herb Boyer and Stanley Cohen were among the first to recognize the potential of these tools and became the first "genetic engineers." In the early 1970s, they began a collaboration that would define the procedures for constructing unique DNA molecules. These procedures made it possible to isolate a gene of interest (for example, the gene for insulin) from a human cell, attach it to a plasmid, and produce thousands to millions of copies of the gene in tiny bacterial factories—a feat that humans had never performed before. See Appendix B for details.

Biotechnology

Scientists recognized the enormous promise of recombinant DNA techniques for yielding large quantities of scarce proteins for scientific study. And entrepreneurs, venture capitalists, and others in the business world saw the enormous potential of producing medically important proteins for therapeutic use in humans.

It didn't take long for scientists to attempt to clone genes from humans and other species in bacteria, although the road was fraught with controversy (see the following sidebar). Crucial to cloning genes of higher organisms was the development of techniques in the 1970s for determining the order of nucleotides in DNA, termed *DNA sequencing*. Scientists needed to know the precise sequences of the DNA molecules they were working with to combine them into novel molecules.

Armed with the ability to determine the order of nucleotides, scientists began studying the structure of genes. In 1977, they discovered that, at least in some species, the DNA in genes is not contiguous. Instead, it occurs in chunks of coding sequence called *exons* that are separated by noncoding sequences termed *intervening sequences*, or *introns*. Introns were also referred to as "junk DNA" because they did not code for protein and, therefore, were thought to have no function. Introns were discovered in all kinds of higher organisms, including humans, and genes containing them, so-called "split genes," appear to be the norm instead of the exception.[11]

Scientists policing themselves

Some scientists recognized early on that recombinant DNA techniques could create organisms with dangerous new traits. After all, the lowly *E. coli* bacterium, the workhorse of the biotech lab, also lives peaceably in the human intestinal tract. Could potentially dangerous cloned genes move from the biotech lab into the human gut via *E.coli*? In 1973, a group of scientists, concerned about the possibility of unwanted genes "escaping" from the laboratory, petitioned the National Academy of Sciences to investigate the dangers of recombinant DNA technology.

A year later, the so-called "Moratorium" letter was signed by many scientists working with the new biotech techniques and was published in the journal *Science*. It called for a moratorium on recombinant work by all scientists worldwide until the safety of this work could be determined. Were the scientists being prudent or overly cautious? At the Asilomar conference in California in 1975, they hashed out all the safety issues, trying to resolve whether the moratorium should remain indefinitely, which would effectively stifle forever this highly promising field of research. Cool heads prevailed as the scientists concluded that the research could continue, provided that weakened strains of bacteria were developed that could not survive outside the lab. They also called for elaborate containment facilities for research on the most dangerous genes. During these highly charged meetings among the scientists, the media fanned the flames in the background, raising the public's fears. However, self-regulation, government oversight, the use of weakened strains, effective containment methods, and the eventual commercialization of many biotech products without incident have helped allay fears of a deadly mutant microbe that is out of human's control, as portrayed in the 1969 novel *The Andromeda Strain* by Michael Crichton.

The use of recombinant DNA techniques to produce genetically modified organisms (GMOs) also has revolutionized the agricultural industry. Wheat, corn, and soybeans have been modified to be pest, disease, and herbicide resistant. Pharmaceuticals are also being developed in GMO crops, such as Vitamin A-boosted golden rice, and bananas engineered to produce vaccines.

Cloning the human insulin gene in bacteria was one of biotechnology's first successes. Surprisingly, it did not occur in a university lab, but in one of the world's first biotechnology companies, Biogen. This company was launched in 1978 on the heels of the first biotech company, Genentech, which was also tackling the cloning of insulin but by a different method. Since then, biotechnology has spawned thousands of new start-up companies. And for the first time in history, biologists left the ivory tower in great numbers to become founders and active participants in this exciting new industry. The founding of Genentech, a partnership between entrepreneur Bob Swanson and scientist Herb Boyer, is the stuff of legends. Genentech, now part of Roche, is one of the most successful biotech companies to date.

Human insulin was a very good pick for these first cloning endeavors. Previously, diabetics had to inject insulin derived from cows and pigs. Although such animal insulin was very close to the human protein, differing by only a few amino acids, that difference caused an allergic response in some patients whose immune systems saw the animal protein as foreign. Having human insulin would solve the immunological problem. Also, the worldwide market for insulin was enormous, with tens of millions of diabetics requiring insulin shots to survive. The new biotech start-ups also got down to the business of cloning other, similarly attractive large-market proteins such as interferon, human growth hormone, and erythropoietin to boost red blood cells in anemic patients. Unlike insulin, many human proteins have complex structures containing other molecules, such as sugars, in addition to amino acids. These proteins could not be produced in bacteria because bacteria lack the machinery to handle the additional structures. To solve this problem many valuable human proteins are commercially produced in yeast and mammalian cells, as well as other cell types.

Unraveling the code

James Watson of double-helix fame was awed by the Human Genome Project (HGP): "Had anyone suggested in 1953 (the year the double helix structure was published) that the entire human genome would be sequenced within fifty years, Crick and I would have laughed and bought them another drink."[12] In 1988, Watson was asked to head up the fledging genome project, with the goal of determining the order of the approximately 3.2 billion to 3.4 billion nucleotides (containing base pairs of A's, T's, C's, and G's) that make up DNA in the human genome. The term *genome* refers to the entire complement of genetic information located in the nucleus of every cell in a given species; *genomics* is the study of the genome. Initially, the HGP was a collaborative effort of universities and research centers in the USA, Europe, and Japan, and was mostly funded by governments.

By 1998, a second, private effort was launched by a new company, Celera Genomics, led by Craig Venter. He was determined to sequence the genome and sell the newly discovered gene sequences to pharmaceutical companies for commercial use. This was in sharp contrast to the public effort which intended to provide the sequences free of charge to other scientists. The ensuing rivalry was often played out in the media and led to a race to the finish line. In the end, science benefited because two sequences were derived by different means to compare to one another. On June 26, 2000, President Bill Clinton and Prime Minister Tony Blair of Great Britain announced the completion of the first draft of the HGP, with Venter representing the private effort and Francis Collins (who replaced Watson) from the public effort standing both at Clinton's side. A truce appeared to have been reached, and both the public and private efforts received equal credit for this tremendous accomplishment. As Clinton remarked, "Today we are learning the language in which God created life."[13]

Finding a way to scale up the quantity of DNA for sequencing was critical to the HGP. In the early days of biotech, cloning gene segments in bacteria was the method of choice. However, *polymerase chain reaction* (PCR) was a much better technique for amplifying a particular sequence of DNA that might only be present in a minute amount. It was invented by Kary Mullis, who received the Nobel Prize for this work in 1993. In addition to PCR, highly automated *sequencing machines* were developed to rapidly read the order of nucleotides in segments from the genome. Powerful computer programs entailing hundreds of computers could arrange these discrete segments and assemble them into the final genome sequence.

After the first draft of the human genome was announced with much fanfare in 2000, a so-called "finished version" became available in 2004. With each segment having been sequenced up to ten times, this version was deemed to be 99.9% accurate. The number of actual protein-coding genes has been revised downward from earlier estimates to around 25,000. This makes the human genome about equivalent in coding sequences to that of the mouse and chicken, with the sobering realization that we actually have fewer coding sequences than the mustard plant or ordinary rice.[14] Appendix C, "Complexity of the Genome," further discusses the human genome.

Conclusion

This chapter explained that the first wave of progress in preventing premature death started with astute observations about how diseases are spread. The gradual adoption of hygienic practices and epidemiological protocols in hospitals and communities saved countless lives. The discoveries of vaccines and antibiotics started a second wave of disease prevention. These breakthroughs had a dramatic impact on mortality from infectious diseases, especially in the developed world. The third wave of breakthroughs is coming from cell and molecular biology. The sequencing of the human genome and development of novel biotechnology tools is transforming medicine. The cloning of therapeutic proteins, the creation of monoclonal antibodies and other breakthroughs may eventually cure heart disease, diabetes, and cancer. Building on the biomedical foundations laid thus far, the next chapter presents a snapshot of promising new bioscience technologies, and how they are beginning to deliver on the promise of novel diagnostics, treatments, and cures.

Endnotes

[1] http://en.wikipedia.org/wiki/Bubonic_plague.

[2] Robert N. Butler, *The Longevity Revolution: The Benefits and Challenges of Living a Long Life* (New York: Public Affairs, 2008).

[3] *Ibid.*

[4] Charles E. Rosenberg, *The Cholera Years* (Chicago: University of Chicago Press, 1987).

[5] B. Guyer, M. A. Freedman, D. M. Strobino, E. J. Sondik, *Pediatrics*, no. 6 (2000): 1307–1317.

⁶*The Top Ten Causes of Death*, World Health Organization: Fact Sheet No. 310 (November 2008).

⁷The term *vaccination* for the procedure and the term *vaccine* for the material injected were named after *vacca*, the Latin term for a cow.

⁸Gerald Karp, *Cell and Molecular Biology: Concepts and Experiments* (Hoboken, NJ: John Wiley & Sons, 2008).

⁹Nucleotides in DNA contain a deoxyribose sugar, a phosphate group, and one of four different nitrogen-containing chemical bases: adenine (A), thymine (T), guanine (G), and cytosine (C) which show up in pairs of AT and CG (refer to Figure 2.1).

¹⁰Cloning also refers to making copies of cells or whole organisms such as animals and plants. Refer to Figure 3.1 in Chapter 3, "Snapshot of the Biosciences."

¹¹Introns must be removed from mRNA before protein can be made (see Appendix A for a description of mRNA). After mRNA is transcribed from DNA in the nucleus, introns are cut out—a process called splicing—leaving only intact coding sequences referred to as exons. It proved a challenging task to directly clone human genes in bacteria because bacterial cells have none of the splicing enzymes that could remove introns. However, scientists Howard Temin and David Baltimore discovered that a group of viruses that have RNA as their genome possess an enzyme that can make a DNA copy of this RNA. The enzyme was termed *reverse transcriptase* because it reverses what was considered the normal flow of genetic information from DNA to RNA. Scientists who were attempting the first cloning of human genes in bacteria realized this enzyme could provide the means of solving the intron problem. They would isolate the mRNA of their target gene, such as human insulin. The mRNA is free of introns and, therefore, would provide a clean, uninterrupted template of the insulin gene. The mRNA was treated with reverse transcriptase to form a DNA copy of the gene that could be inserted into a plasmid for cloning. Bacteria that were transformed with the hybrid plasmid would churn out huge quantities of human insulin.

¹²James D. Watson, *DNA* (New York: Alfred A. Knopf, 2003).

¹³"Text of Remarks on the Completion of the First Survey of the Entire Human Genome Project," White House Press Release (2000). Available at http://Clinton5.nara.gov/WH/New/html/genome-20000626.html.

¹⁴Gerald Karp, *Cell and Molecular Biology* (Hoboken, NJ: John Wiley & Sons, 2008).

3

Snapshot of the biosciences

The scope of the biosciences is very wide, encompassing the basic sciences, technologies, and methods used in research, as well as the industries that apply and commercialize the novel findings of bioscientists.[1] Basic bioscience includes all disciplines in the natural sciences that relate to living organisms, such as cell biology, genetics, zoology, and botany. Applied bioscience includes such disciplines as biotechnology, medicine, pharmaceuticals, agriculture, and energy, as well as related software and information systems developed for biological applications.

Numerous industries are or will be affected by the biosciences, such as insurance (genetic screening), personal security (biometrics such as DNA identification), family counseling (genetic testing for hereditary conditions), and, of course, healthcare (gene therapies). Given the wide footprint of the biosciences, we examine here the main bioscience technologies that are important in medicine. Table 3.1 focuses on cellular and subcellular technologies; Table 3.2, later in the chapter, lists technologies based more on chemistry, engineering, materials, and information technology. Chapter 4, "Bio-driven Convergence," covers these latter technologies in more detail.

DNA-based technologies

Let's look at some of the major technology platforms whose primary focus is DNA. Thanks to the Human Genome Project and other research efforts, we know a great deal more now about the makeup of human DNA and its functions. Our review will be short and offers no more than a snapshot of a complex, dynamic and continually changing set of technologies and scientific disciplines.

Table 3.1 Important biology-based technologies

Area of Focus	Technology or Platform	Main Purpose	Sample Application
DNA	Genomics	Mixed	Conducting basic research on heart disease and cancer
	Genetic testing	Diagnostic	Testing for sickle cell anemia or Tay-Sachs
	DNA chips	Diagnostic	Determining genes activated in lymphoma cancers
	Pharmacogenetics	Therapeutic	Determining customized drug dosing
	Recombinant DNA	Therapeutic	Enabling large-scale production of insulin
	Gene therapy	Therapeutic	Fixing a gene defect to prevent blindness
	DNA fingerprints	Forensics	Solving crime or determining paternity
RNA	Antisense	Therapeutic	Preventing blindness in AIDS patients
	RNA interference	Therapeutic	Preventing viral disease
	Micro RNAs	Mixed	Treating cancer and heart disease
Protein	Vaccines	Mixed	Preventing flu; treating cancer
	Monoclonal antibodies	Mixed	Treating lymphoma and rheumatoid arthritis
	Proteomics	Mixed	Conducting basic research on protein function
	Immunoassays	Diagnostic	Diagnosing infectious diseases such as AIDS
Cell/Other	Stem cells	Therapeutic	Repairing damaged heart tissue
	Antimicrobials	Therapeutic	Curing bacterial pneumonia

Genomics

Genomics is the study of the structure and function of genes, with a major focus on identifying genes involved in human diseases. More than 1,000 genes have been linked to specific rare genetic diseases. Initially, studies focused on diseases caused by mutations in a single gene. Researchers analyzed multiple generations of families with a high incidence of the disease, and then scanned their genomes for regions showing similarities. That DNA region was then isolated so that researchers could pinpoint the exact mutation responsible for the disease. Such family studies led to the discovery of the gene mutations responsible for diseases such as Huntington's, Duchenne muscular dystrophy, and cystic fibrosis, among many others. If you have these gene mutations, you will get the diseases largely independent of your environment.[2]

Common diseases such as cancer, heart disease, diabetes, and depression entail multiple mutations and possibly many genes. Environmental factors such as smoking and obesity also contribute to common diseases. As researchers like to say, "Genes load the gun and the environment pulls the trigger." The genes of many common diseases cannot easily be studied through familial linkage, as with rare genetic disorders. Instead, researchers must compare the DNA of people with the disease to the DNA of individuals of similar ethnic background who are not affected, to pinpoint genes that are associated with the disease. Scientists do this by exploring variations in the human genome called *single nucleotide polymorphisms* (SNPs), as explained in Appendix C, "Complexity of the Genome."

Genetic testing

This technology involves analyzing chromosomes (DNA), RNA, proteins, and certain metabolites to assess a person's propensity to develop a genetic disease or to pass it on to offspring. More than 1,200 genetic tests are currently available for a range of genetic diseases (see accompanying sidebar), with more being developed. Genetic tests are typically conducted on samples of blood, hair, skin, amniotic fluid (around the fetus in utero), semen (in rape cases), or other tissues or fluids.

Types of genetic tests

- **Prenatal**—These tests are conducted during pregnancy to determine the unborn child's risk for particular genetic diseases, such as spina bifida and Down syndrome. They can help a couple determine whether to abort the pregnancy or prepare for illness.

- **Newborn**—This routine screening of newborn infants is the most common application of genetic testing. Genetic abnormalities such as phenylketonuria and hypothyroidism can quickly be detected and treated in babies.

- **Presymptomatic**—This testing can determine whether a person is at risk of developing certain genetic diseases later in life, especially when family members have the disease. An example is testing for the BRCA1 mutation, which greatly increases risk of breast cancer.

- **Carrier**—This testing determines whether two partners have a higher risk of passing on the genes for sickle cell anemia, Tay-Sachs, or other genetic diseases to their offspring.

- **Diagnostic**—These tests are performed if an individual has symptoms of a genetic disease, such as cancer, to confirm its specific diagnosis.

- **Forensic**—This testing can determine whether an individual is a victim or perpetrator of a crime, or whether a biological relationship exists between individuals (for example, to establish paternity).

- **Direct-to-Consumer**—These tests can be obtained from the Internet and give your risk for various diseases.

Recent findings from the Human Genome Project enable scientists to cast a wider net in determining the role of genetic heritage. Researchers have sequenced the entire genome of some individuals, including a few individuals who were willing to pay as much as $350,000 for the privilege. Private companies now offer this service, mostly to the very rich, without any immediate or proven benefits.[3] However, technology is moving fast, and the federal government is supporting research to reach the goal of a $1,000 genome in the next decade. In the near future, we might each possess a microchip containing our full genome sequence, plus our individual probabilities of developing major diseases.

Currently, you can buy a less comprehensive glimpse of your genetic destiny for about $400. Companies such as 23 and Me, deCode Genetics, and Navigenics can analyze a bodily sample, such as a cheek scraping, for several hundred thousand SNPs distributed across your chromosomes. This can determine your risk profile for about 20 common diseases, including Alzheimer's disease, breast cancer, rheumatoid arthritis, and multiple sclerosis. They also give probabilities for a sprinkling of physical traits, such as the ability to taste bitter flavors or your tendency to gain weight, and information about what part of the world your ancestors came from.[4] A problem with these tests is that the common diseases they cover have multiple genetic causes, many of which aren't known, including the role of environmental causes, leading to an incomplete or misleading picture of one's actual risk. A further concern is that many people will unnecessarily worry and seek useless follow-up tests (at great cost) when they know they have even a small probability of disease. These genetic tests represent a growing trend in "direct-to-consumer" testing, in which the consumer obtains a test over the Internet without having to go through a doctor. Hundreds of these tests are on the market, but the federal government and most states have been slow to take regulatory action, leaving consumers on their own when it comes to judging the efficacy and reliability of the results.

DNA chips

DNA chips, also called *DNA microarrays*, enable scientists to monitor the expression of thousands of genes at one time from a given tissue across varying conditions. Most physiological processes in the cell respond to changes in the environment by changing the level of expression of genes involved in those processes. Some genes become more active, leading to more transcription and translation of their protein product, and others might turn off altogether. For example, changes in expression of many genes, sometimes hundreds or thousands, can occur as a cell transforms from a normal to a malignant state. DNA chips can detect which genes turn off and on, and to what degree, all at the same time. So instead of looking at the changes in a disease process one gene at a time, scientists can observe changes to a whole set of genes at once.

In 1996, Stephen Fodor of Affymetrix Inc., developed the idea of seeding glass chips with DNA, similar to how transistors are put onto silicon chips. Droplets containing millions of copies of the same DNA fragment (from a specific gene) are precisely located on the glass grid in microscopic wells. DNAs from thousands of different genes can be applied to each grid. In one type of procedure, mRNAs are extracted from sample tissues and are then attached to fluorescent tags. The tagged molecules are applied in solution to the surface of the glass grid and allowed to react (via complementary base pairing) with the attached DNA molecules, representing thousands of human genes. Each spot on the chip is then analyzed to measure the concentration of the fluorescent DNA. The more fluorescent the spot, the greater the expression of the corresponding gene. In a typical experiment, scientists analyze the differences in gene activity (expression) between diseased and normal cells so as to identify genes involved in the disease process.

Today microarrays are used mainly in research, but they hold great promise for improving the diagnosis of diseases such as lymphoma cancer, which can be of numerous types that are difficult to distinguish. They can also be used to improve the dosing of prescription drugs. The AmpliChip CYP450 Test, made by Roche laboratories, is the first microarray approved by the FDA for diagnostic use in the United States. It is a DNA chip that doctors can use to help determine the correct dosing of SSRIs (antidepressant drugs) such as Prozac and Paxil.[5]

Pharmacogenetics

As more personal genetic information becomes available about patients, prescribing drugs will no longer be based on "one size fits all." Instead, drugs will be tailored to the individual. Choosing a drug based on a patient's genetic profile is called *pharmacogenetics*. This process is important because most drugs are effective for only about half the individuals being treated. For example, the leading breast cancer drug, Herceptin, is effective in only the 20% to 30% of patients whose tumors express high levels of the Her2 protein.[6] But doctors usually don't find out which patients it works for until treatment is well under way and the effects become visible. Likewise, the colon cancer drugs Erbitux and Vectibix won't work for 40% of patients whose tumors have a specific gene mutation. Because these drugs can cost up to $10,000 per month, it is imperative to screen patients genetically before embarking on such costly treatments.[7]

Pharmacogenetics is also starting to play a role with widely prescribed antidepressants such as Prozac and Paxil. The cytochrome P450 (CYP450) enzymes in the liver metabolize these drugs. Genetic variations in the genes coding for these enzymes affect how quickly these drugs are broken down. Some people metabolize slowly, potentially causing these drugs to accumulate to toxic levels in the body; others metabolize too rapidly for the drugs to have the expected effect. Knowing which variants of these enzymes a patient has should help doctors prescribe the appropriate dosage of antidepressants.[8] Patients taking warfarin (Coumadin) to prevent blood clots also would greatly benefit from genetic testing because two specific genetic markers could help doctors determine the appropriate dosage of this medication. This would be a major improvement in care because thousands of patients are hospitalized each year due to internal bleeding from high doses or blood clots from low doses.[9]

Cloning

Cloning is the process of creating multiple copies of a gene, a cell, or a whole organism. In the most common type of cloning, a target gene is produced in multiple copies using recombinant DNA (rDNA) techniques. The gene of interest can reside in any living organism, from a virus to human cells. Bacteria, yeast, or mammalian cells can be used as small biological factories, churning out many copies of the target gene.

The first generation of cloning focused on making substitutes for natural proteins, such as the hormone insulin, which was expensive to extract from animals and purify. Safety concerns surfaced about some therapeutic proteins obtained from human cadavers, making cloning in bacteria very attractive. For example, in the 1980s, scientists discovered that the infectious agent that causes a fatal neurological disease, Creutzfeldt-Jakob disease (CJD), contaminated the human growth hormone used for treating dwarfism. About 30 people treated with the cadaver-derived hormone died of CJD, which was unknowingly transmitted to these patients. Producing proteins in bacteria avoids any risk of infecting unsuspecting patients with dangerous viruses or other agents from donor tissues.

The second generation of cloned molecules focused on treatments for cancer, immune disorders, and infectious diseases. These
molecules consist of not only hormones, but also growth factors,
enzymes, and cytokines (proteins secreted by immune system cells
that alter the behavior of other immune cells). They have now also
been successfully cloned in bacterial cells, yeast, and mammalian
cells. Appendix B, "Cloning Genes," explains how a typical cloning
experiment is performed.

A third generation of cloning made the national news in 1996,
with the birth of Dolly the sheep at the Roslin Institute in Scotland.
Dolly was the first mammal cloned from an adult cell. She was created from an udder cell taken from a six-year-old donor sheep. Scientists removed the udder cell's nucleus and transplanted it into the egg
from a second sheep (whose own nucleus was removed). They then
transplanted this egg into the uterus of a third sheep (the surrogate
mother) to grow and develop into Dolly, just as a normal embryo
would (see Figure 3.1). This experiment refuted the previous dogma
that cells couldn't be returned to their undifferentiated state after
they had become specialized in particular tissues in adult mammals.
The udder cell nucleus from the adult sheep was somehow reprogrammed by the egg to yield a viable embryo that could grow and
produce all the cell types found in the adult. This technology was
termed *somatic cell nuclear transfer (SCNT)*.

The Roslin team headed by Ian Wilmut was not interested in
cloning humans. Their aim was to find a scalable method for introducing desired genetic changes into livestock. It took the Roslin team
277 attempts before they had a viable embryo that produced Dolly.
This biological tour de force led to an explosion of interest in the
methodology as scientists would like to take a patient's own cell,
transplant the nucleus into an unfertilized human egg, and produce
an embryo from which stem cells (see the "Stem Cells" section later
in this chapter) could be harvested to treat a whole range of human
maladies. This has been dubbed *therapeutic cloning*, in contrast to
reproductive cloning, in which the embryos (derived from SCNT) are
created with the purpose of producing multiple, identical copies of
humans. The movie *The Boys from Brazil* in which a clone of
young Hitlers is produced, represents a nightmare scenario for this
technology.[10]

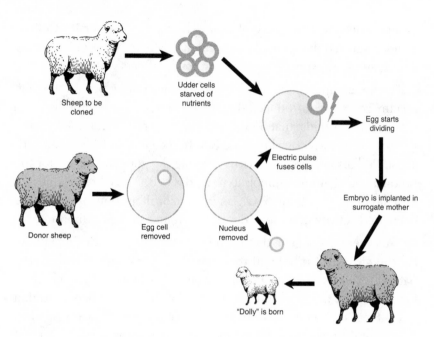

Figure 3.1 Cloning of Dolly

Gene therapy

Scientists have identified numerous gene defects as likely causes of thousands of medical conditions, ranging from Alzheimer's and Tay-Sachs disease to cancer and diabetes. The ultimate cure for such conditions is to replace the defective gene with a healthy copy, a technique known as *gene therapy*. Since its beginnings in the early 1990s, when it was heralded as "the medicine of the future," gene therapy has seen many ups and downs. Gene therapy has cured children with SCID, a severe combined immunodeficiency (the boy-in-the-bubble syndrome). But it also resulted in the death of 18-year-old Jesse Gelsinger due to a severe immune reaction to his therapy, plus multiple other deaths from leukemia related to gene therapy treatments.

Although the FDA has not yet approved any gene therapy treatments, hundreds of trials are in various stages of progress. One recent success is a trial to restore sight to those who are blind. Dozens of genes have been identified for inherited eye diseases. One such disorder, Leber congenital amaurosis 2, affects about 3,000 American children. No treatment previously existed, and most of those affected are completely blind by the age of 40. In 2008, independent research groups from the University of Pennsylvania and University College London showed that gene therapy can partially restore sight to young adults, and they expect even more profound effects with children.[11]

Gene therapy treatments generally fall into two types. In the first, healthy genes are introduced directly into target areas of the body via viruses or plasmids that have had their own DNA removed and replaced with corrective human DNA. In recent promising clinical trials on Parkinson's disease, a virus carrying the gene that codes for the production of gamma-aminobutyric acid (GABA) was directly injected into the brain of patients.[12] GABA is a major neurotransmitter whose decreased activity in these patients contributes to their symptoms.

Another type of gene therapy method involves repairing a patient's cells in vitro (test tube) and then returning the repaired cells to the patient (see Figure 3.2). Doctors have treated more than 30 children for SCID this way, and more than 90% of them have been cured, a dramatic improvement over the traditional bone marrow transplants. But several SCID patients developed leukemia as a side effect of their gene therapy, illustrating the risks involved with these lifesaving treatments.[13]

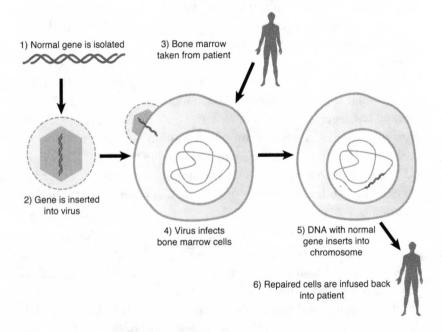

1) Normal gene is isolated

2) Gene is inserted into virus

3) Bone marrow taken from patient

4) Virus infects bone marrow cells

5) DNA with normal gene inserts into chromosome

6) Repaired cells are infused back into patient

Figure 3.2   Gene therapy procedure

Although gene therapy shows great promise for a variety of diseases, the most intense focus is on treating cancer. Gene therapy can combat cancer from a variety of different angles. One strategy is to modify defective genes involved in the cancer process. For example, human cells have a tumor-suppressor gene, called p53, that normally prevents cells from becoming cancerous. Several types of cancer have been linked to a defective p53 gene, making this an attractive target for gene therapy. Another strategy is to make cancer cells more recognizable to the immune system, or to make them more vulnerable to chemotherapy and radiation by bringing in genes that heighten their sensitivity.[14]

RNA-based technologies

DNA is the carrier of our genetic code, whereas RNA helps translate the code into proteins. Scientists are exploring RNA's various roles in the cell in the hope of producing novel treatments for human disease. We describe two such technologies below.

Antisense technology

This is essentially a gene-silencing technology, called "antisense" because it employs a strand of nucleotides that is complementary to the target gene's messenger (the "sense" strand). When the genetic sequence of a particular gene is known, scientists can construct a complementary RNA sequence that binds to the mRNA from that gene, creating a double-stranded molecule that cannot be translated into protein.

Drug companies are developing antisense technologies to treat diseases such as cancer, diabetes, arthritis, and Duchenne muscular dystrophy. Isis Pharmaceuticals received FDA approval in 1998 for the first antisense drug, Vitravene. Doctors have used the drug to treat cytomegaly retinitis, a retinal disease that occurs in patients afflicted with AIDS that can lead to blindness if left untreated.[15] One of the early successes of antisense technology was the creation of the Flavr Savr tomato by a California company, Calgene. Traditionally, tomatoes are harvested green to avoid ripening too soon, becoming soft, and rotting on long hauls. Upon reaching the market, green tomatoes are then treated with ethylene gas to accelerate ripening, at the expense of the delicious flavor of a locally grown tomato. Calgene's technology targeted the enzyme that softens the tomato's cell wall. In treated tomatoes, a strand of antisense RNA binds to the softening enzyme's messenger RNA and prevents its translation to active enzyme. The cell wall of the fruit does not go soft as quickly, so Flavr Savr tomatoes could be allowed to ripen on the vine for a better flavor and then be shipped. This was the first commercially grown genetically modified food that the FDA granted a license for human consumption.

RNA interference (RNAi)

This is also gene-silencing technology, but it's based on short double-stranded RNAs called short-interfering RNAs (siRNAs). When added to a cell, these small molecules can bind to a segment of mRNA (from the gene of interest) that has a complementary sequence. This binding causes target mRNA to be cut up, preventing it from being translated into protein. RNAi was discovered in 1998 and is known to occur naturally in worms, plants, and mammalian cells.

RNAi has become a hot area for drug research and is the main focus of several biotech companies. They design short segments of double-stranded RNA that are complementary to the mRNAs of genes in humans or infectious agents. For example, Alnylam Pharmaceuticals has created a promising RNAi therapeutic drug for respiratory syncytial virus, which causes more than 17,000 deaths per year. The drug silences one of the virus's genes, rendering it incapable of causing infection.[16] The biggest challenge for an RNAi-based drug is getting it to the site where it is needed in the body because double-stranded RNA looks like a virus to the cell and can be attacked. Also, enzymes in the bloodstream can destroy these RNA molecules.

Protein-based technologies

Proteins are organic compounds composed of amino acids, arranged in linear chains, and held together by peptide bonds. They make up many of the structural elements of cells and are the primary component of the technologies described below.

Vaccines

When a human is infected with a pathogen, such as the measles virus or the tetanus bacteria, the immune system recognizes components of the microbe as foreign or "nonself." These foreign components are called *antigens*. Once detected, cells called lymphocytes spring into action, resulting in the production of (1) antibodies that can neutralize antigens, (2) killer lymphocytes that can destroy foreign cells, and (3) memory cells that "remember" the antigens. In case of future exposure to the same pathogen, these memory cells help quickly recognize and destroy it, preventing the infected individual from becoming sick.

Vaccination (also referred to as immunization) is the process of tricking the body into building up a resistance to a specific disease agent, as described above, without actually causing illness. Traditionally, vaccines have consisted of either killed pathogens or attenuated (weakened) ones. The Salk polio vaccine is an example of a killed vaccine. In contrast, the Sabin oral polio vaccine as well as the measles, mumps, and rubella combination vaccine (known as MMR), are *attenuated* vaccines. They contain active (live) pathogens whose virulence has been sufficiently weakened to avoid illness while still eliciting a strong immunological response to prevent future disease. The current trend in vaccines is to use purified antigenic components from the pathogen, rather than whole cells. Examples are the tetanus and diphtheria vaccines. These are composed of inactivated protein toxins, called *toxoids*, that are derived from the pathogen. These toxins, which are secreted by the infecting pathogen, are responsible for the disease symptoms.

Two of the newer approaches to vaccine technology, both still experimental, are *DNA vaccines* and *edible vaccines*. DNA vaccines consist of the genes for antigenic components of pathogens that can be injected into the patient's muscle by using a "gene gun."[17] Edible vaccines are cloned into traditional crops such as bananas and potatoes to simplify delivery and acceptance by those who need immunization. They hold promise for developing countries where traditional vaccines are difficult to transport, store, and deliver.[18]

Monoclonal Antibodies

As explained in the previous chapter, this technology originated in 1975 in order to fight foreign invaders as well as tumor cells. For example, by fusing antibody-producing B lymphocytes from mice with mouse myeloma (tumor) cells, a *hybridoma* was produced that could churn out identical antibodies of a single specificity (mono-clonal). However, the problem with these early mouse-derived mono-clonal antibodies (MAbs) was that the human immune system would see them as foreign, resulting in an immune response for some patients. Any such immune response can reduce the efficacy of future treatments and in some cases can result in allergic reactions that are life threatening to some patients. Thus, scientists sought to modify MAbs to make them less mouse-like and more human. Over time, the antibody structure changed from 100% mouse protein to *chimeric* MAbs (approximately 66% human protein, 33% mouse protein), to *humanized* MAbs (approximately 90-95% human protein), and now to fully human MAbs (100% human protein).[19]

In addition to be being less antigenic, these fully human MAbs often have higher affinities for their targets than do chimeric or humanized antibodies. One of the most promising technologies for producing fully human MAbs involves the use of transgenic mice in which the mouse antibody genes have been inactivated and replaced with human antibody genes. A research collaboration between the Kirin Brewery Company and Medarex, Inc. created one such mouse system. This so-called KM mouse can be immunized with specific antigens—for example, tumor proteins—to produce human MAbs against those very antigens. The Danish company Genmab uses a similar transgenic mouse technology from Medarex Inc. to develop fully human antibodies for various types of cancer. Early clinical data is very promising in terms of efficacy while also eliciting less allergic responses or other side effects than previous generations of antibody products containing mouse or other animal proteins.[20] Newer tech-nologies deploy just antibody fragments rather than whole antibodies. Potential benefits of these smaller molecules are that they can pene-trate inaccessible tissues such as tumors more deeply, tend to distrib-ute more evenly across tumors, and can be cleared from the body more quickly, which means less toxicity for the patient.

Today hundreds of biotech companies focus on monoclonal antibody-based therapies. At least 21 drugs on the market are based on these specialized molecules, and another 38–40 are in clinical trials. The top five monoclonal antibody drugs—Avastin (cancer), Herceptin (breast cancer), Humira (autoimmune diseases), Remicade (autoimmune diseases), and Rituxan (cancer)—account for 80% of the market.[21] Monoclonal antibodies (MAbs) are also used in diagnostics. For example, MAbs are used to detect human chorionic gonadotropin (HCG) in home pregnancy tests. Doctors also use MAbs in diagnostic tests to detect the presence of HIV (AIDS virus) or hepatitis virus in infected individuals. Coupled with cytotoxic drugs or radioactive chemicals (which kill tumor cells), MAbs can act as a magic bullet by delivering the toxic substance directly to the tumor, while reducing the side effects of chemotherapy.[22]

Proteomics

A relatively new term, *proteomics*, is the study of all proteins, expressed in cells, tissues, or entire organisms. The field of proteomics is the next logical extension of the Human Genome Project, which has revealed how many genes are in the human genome. However, the HGP does not reveal the additional proteins created through splicing (cutting) of mRNAs, which greatly increases the size of the human proteome: the full complement of an organism's proteins (see Appendix C, "Complexity of the Genome," for additional discussion). In addition to knowing the amino acid sequence, scientists want to understand a protein's three-dimensional structure, its interaction with other proteins, its association with a particular disease, and, ultimately, its overall functioning.

Recent advances in high-speed computing, two-dimensional gel electrophoresis, and mass spectroscopy permit (1) viewing the entire protein profile of a particular tissue under varying conditions, (2) separating proteins from one another, and (3) analyzing each protein's structure. Protein microarrays, also called *protein chips*, enable scientists to study interactions of proteins in complex mixtures under a variety of circumstances. In addition, identifying fluctuations in proteins throughout a disease process can reveal *biomarkers*, or signs of the disease.[23] Many researchers are calling for a human proteome project (similar to the Human Genome Project) to survey all the proteins present in human tissues.

Cell-based and other technologies

This final section first discusses *stem cells*, which are undifferentiated (immature) cells found in the embryo as well as in of
adult tissues. Stem cells have tremendous therapeutic promise
for curing diseases such as diabetes, heart disease, and Parkinson's disease. We then discuss *antimicrobials*, which are mostly
small molecules produced by living organisms or synthesized in
the laboratory. They have cured many of the most feared diseases but are losing effectiveness due to growing microbial
resistance.

Stem cells

At the end of 2008, the FDA gave approval for the first trial of a stem
cell treatment in humans to the Geron Corporation.[24] It will treat
patients paralyzed by spinal cord injury with an injection of glial cells
derived from embryonic stem cells. Scientists hope that these cells
will restore damaged or destroyed nerve cells and enable patients to
regain use of their limbs. Even if this trial is successful, it will likely
take years before these types of treatments are broadly approved and
become widely available to patients in need.

Embryonic stem cells have the capacity to evolve into any type of
tissue in the human body (nerve cells, muscle cells, and so on) and
have therefore been termed *pluripotent*, or having more than one
potential end state. They have sparked great enthusiasm for their
promise to regenerate diseased tissues, as well as major moral controversy because they use human embryos. Stem cells appear in the
early embryo within a week after conception and have the ability to
transform into more than 200 cell types that make up all the tissues of
the human body.

A major breakthrough in 2006 showed that ordinary mouse cells
could be reprogrammed to a pluripotent stem cell–like state without
employing embryos. These cells have been termed *induced pluripotent stem cells (iPS)*. Many scientists have switched to this new line of
research because it appears to be a much faster and less ethically
controversial way to obtain patient-specific stem cells for human
treatments.[25]

Adult stem cells are a special population of cells within organs and tissues of the body that normally replace dead or injured cells. For example, when dead skin cells are sloughed off, adult stem cells in the skin create replacement cells. Although they don't have the same degree of plasticity as embryonic stem cells, scientists are attempting to grow adult stem cells in culture, for the purpose of repopulating diseased or injured tissues with healthy cells. In the future, they might be used to provide insulin-producing cells in diabetics, repair damaged heart muscle after a heart attack, or replace dopamine-producing cells in Parkinson's patients.[26]

Antimicrobials

Infectious diseases are still one of the leading causes of death worldwide. In addition to well-known pathogens that have been with us for ages, new ones are continually emerging due to mutation and natural selection. Since the 1970s, more than 30 new disease agents have been identified, including E. coli 0157, HIV, SARS, and the swine H1N1 influenza virus that emerged in the spring of 2009. Scientists are also concerned that many of the most effective antibiotics, the wonder drugs of the twentieth century, are becoming ineffective due to bacterial resistance (see the following sidebar).

The major antimicrobial agents are antibiotics that inhibit or kill bacteria, and antivirals that inactivate viruses. Pharmaceutical companies are continuing their hunt for an antimicrobial that will be the next miracle drug, such as penicillin. They search for novel microbes and plant or animal specimens from both land and marine environments worldwide to find small molecules that will be effective against pathogens. By using *high-throughput screening*, scientists can examine thousands of these natural compounds per day for activity against microbial targets using highly automated robotic systems. Pharmaceutical companies are pursuing other routes as well, such as *rational drug design*. They are designing novel small molecule drugs that can bind to specific microbial cell targets to kill the microbes or prevent their multiplication. Also, genomics opens up the prospect of finding previously unknown genes that are unique to pathogens and can become new targets for antimicrobials.

Revenge of the superbugs

Remember the days when your family doctor would prescribe an antibiotic to treat that upper respiratory infection without knowing for sure if you had a bacterium or virus? Antibiotics, once miracle drugs that cured some of the most dreaded diseases of mankind, have been overprescribed to such a degree that many bacteria have become resistant to their effects. The resistance stems from plasmids that carry antibiotic resistance genes that destroy antibiotics. These plasmids pass between bacteria via a mating-like process called conjugation. In environments where antibiotic is present (such as in a humans receiving antibiotic therapy) bacteria that carry plasmids with antibiotic-resistance genes are selected for and become the prevalent type.

The CDC reported that in 2005 nearly 94,500 people developed a serious infection of a virulent *Staphylococcus aureus* strain that is resistant to the antibiotic, methicillin, one of the last antibiotics that can effectively treat this infection. Approximately 19,000 of these patients died from these infections. Termed MRSA (for methicillin-resistant *Staph. aureus*), this bacterium has become the scourge of hospital wards as well as locker rooms. About 85% of all MRSA infections emanate from a healthcare environment. The source of infection is a patient who has an active MRSA infection or one who carries the bacteria but has no symptoms. It spreads most commonly by hospital staff who transfer the bacteria via their unwashed hands.

Conclusion

We have examined just some of the remarkable technology and science platforms that comprise the biosciences today. This chapter focused on biology-based technologies. Table 3.2 lists other bioscience technology platforms, many of which we discuss in later chapters.

Table 3.2 Additional bioscience technologies and platforms

Basic Discipline	Technology or Platform	Main Purpose
Chemistry/Biochemistry	Synthetic biology	Therapeutic
Materials and Devices	Enzymes	Various industrial uses
	Drugs (other than antimicrobials)	Therapeutic
	Artificial and transplanted organs, joints, limbs	Therapeutic
	Xenotransplantation	Therapeutic
	Drug eluting stents	Therapeutic
	Pacemakers, neuroimplants	Therapeutic
	Hearing aids, vision aids	Therapeutic
	Tissue engineering	Therapeutic
Engineering and/or information technology	Remote and Robotic surgery	Therapeutic
	Imaging: MRI, CT, PET, ultrasound	Mixed
	Electronic medical records	Mixed
	Telemedicine	Diagnostic
	Wireless telemetry	Diagnostic
	Supercomputers	Mixed
	Grid computing	Mixed
Multidisciplinary	Systems biology	Mixed
	Bioinformatics	Diagnostic
	Rational drug design	Therapeutic
	Biosensors	Diagnostic
	Nanodevices	Mixed
	Translational medicine	Mixed

The key message is that many promising technologies exist, with new ones emerging yearly. We conclude by highlighting some of the most remarkable accomplishments of the past few decades in the biosciences, with much more to come in the next few decades.

- Researchers have sequenced a variety of genomes from humans, chimpanzees, mice, rats, fruit flies, and many bacteria and viruses as well.

- Scientists have cloned various mammals from mice, cows, goats, pigs, and sheep, to horses, deer, dogs, and cats, and they have attempted to clone human embryos for the purpose of stem cell production (therapeutic cloning). However, "reliable" cloning remains elusive, and human cloning is banned in most countries on ethical and moral grounds.

- Doctors have refined the ability to transplant organs from donors to recipients and to grow human skin and other tissues.

- A growing portfolio of technologies exists to treat cancer, diabetes, and many other serious diseases, including the use of highly targeted monoclonal antibodies and cloned human proteins such as insulin.

- An entirely new field of bioinformatics enables the application of information technology to the field of molecular science.

- Bioengineers have created (experimentally) an artificial heart, an artificial pancreas, and a variety of other innovative medical implants, including increasingly smaller pacemakers for heart patients.

- Research has been ongoing for more than a decade for using genes to treat and cure diseases, although the delivery of replacement genes (gene therapy) in humans is still not a mainstream medical treatment.

- Scientists are developing individualized cancer vaccines and pharmaceuticals that hold enormous promise for what is being called "personalized medicine."

- Researchers can grow and manipulate embryonic and adult stem cells to provide therapeutic benefits for many diseases, with the possibility of treating Parkinson's disease and diabetes, or restoring the use of limbs paralyzed from spinal cord injuries.

Endnotes

1The Hope, Hype, and Reality of Genetic Engineering: Remarkable Stories from Agriculture, Industry, Medicine, and the Environment (New York: Oxford University Press, 2004), by John C. Avise, describes the science of genetic engineering and how genetically modified organisms will serve human needs in medicine, agriculture, and the environment. *Biotechnology Unzipped: Promises and Realities* (Washington, D.C.: Joseph Henry Press, 2006), by Eric S.Grace, offers two- to three-page descriptions of the basic science of DNA; the tools and procedures of biotech; applications of the science for medicine, agriculture, and the environment; and potential discoveries from the oceans and trees. *DNA: How the Biotech Revolution Is Changing the Way We Fight Disease* (Amherst, NY: Prometheus Books, 2007), by Frank H. Stephenson, is focused on how the tools of biotech combat common afflictions such as cancer, AIDS, heart disease, and other diseases. *DNA Technology: The Awesome Skill* (San Diego: Academic Press, 2001), by I. Edward Alcamo, covers the basic science of DNA, proteins, genetic engineering, gene therapy, medical forensics, biotech in agriculture, transgenic animals, and the human genome project for readers with limited scientific background.

[2]Gerald Karp, *Cell and Molecular Biology* (Hoboken, NJ: John Wiley & Sons, 2008).

[3]Amy Harmon, "Gene Map Becomes a Luxury Item," *New York Times* (4 March 2008): D1.

[4]Amy Harmon, "Learning My Genome, Learning About Myself," *New York Times* (17 November 2007): A1.

[5]"About the AmpliChip CYP450 Test"(2009). Available at www.amplichip.us/physicians/about the amplichip.php

[6]In this case, the genetic profile of the tumor (instead of the patient) is used to determine treatment.

[7]Andrew Pollack, "Patient's DNA May Be Signal to Tailor Drugs," *New York Times* (30 December 2008): A1.

[8]S. H. Katsanis, G. Javitt, and J. K. Hudson, "A Case Study of Personalized Medicine," *Science* 320 (4 April 2008): 53–54.

[9]Keith J. Winstein, "DNA Tests May Predict Blood Thinner Dosage," *Wall Street Journal* (19 February 2009): D3.

[10]Ian Wilmut and R. Highfield, *After Dolly: The Promise and Perils of Human Cloning* (New York: W. W. Norton & Co., 2006).

[11]Jocelyn Kaiser, "Two Teams Report Progress in Reversing Loss of Sight," *Science* 320 (2 May 2008): 606–607.

[12]Caroline Cassels, "Gene Therapy in Parkinson's Disease Safe, Potentially Effective," *Medscape Medical News*, (2007). Available at www.medscape.com/viewarticle/558751.

[13]Arthur Nienhuis, "How Does Gene Therapy Work?" *Scientific American* (August 2008): 108.

[14]"Gene Therapy: Cancer Research Topic Proves Challenging but Promising," www.mayoclinic.com/health/cancer-research/CA00046.

[15]Reinhard Renneberg, *Biotechnology for Beginners* (Berlin: Springer-Verlag, 2008).

[16]Keith J. Winstein, "RNA Gene Study Shows Efficacy in Human Trial," *Wall Street Journal* (29 February 2008): B4.

[17]The University of Arizona Immunology Tutorials (January 17, 2006). Available at http://microvet.arizona.edu/courses/mic419/Tutorials/vaccines.html.

[18]Rick Ng, *Drugs: From Discovery to Approval* (Hoboken, N.J: John Wiley & Sons, 2004).

[19]Weiner, Louis M., "Fully Human Therapeutic Monoclonal Antibodies," *Journal of Immunotherapy*: Jan/Feb, 2006, Vol 29, p1-9.

[20]http://www.genmab.com/AboutGenmab/AboutGenmab.aspx

[21]Karl Ziegelbauer and David R. Light, "Monoclonal antibody therapeutics: Leading companies to maximize sales and market share" Journal of Commercial Biotechnology (27 November 2007). Available at www.palgrave-journals.com/jcb/journal/v14/n1/full/3050081a.html

[22]Reinhard Renneberg, *Biotechnology for Beginners* (Berlin: Springer-Verlag, 2008).

[23]Robert F. Service, "Proteomics Ponders Prime Time," *Science* 321 (26 September 2008): 1757–1761.

[24]Ron Winslow and Alicia Mundy, "First Human Stem-Cell Trial Gets Approval from the FDA," *Wall Street Journal* (23 December 2008): A12.

[25]Sally Lehrman, "No More Cloning Around," *Scientific American* (August 2008): 100–102.

[26]"Stem Cell Basics: What Are Adult Stem Cells?" The National Institute of Health (2009). Available at http://stemcells.nih.gov/info/basics/basics4.asp.

4

Bio-driven convergence

with Nanda Ramanujam[1]

Because the field of biosciences lies at the intersection of biology and other scientific or technological disciplines, solutions to many important health problems are likely to emerge from a variety of innovative, sometimes seemingly unconnected, technological platforms. We term this special confluence *bio-driven convergence*. Here are some examples:

- Transferring medical data over communications networks (*telemedicine*)
- Using sensors and wireless technology to monitor patients outside the hospital setting or when the patient is mobile in the hospital (*remote diagnostics*)
- Cloning genes in plants and microorganisms to produce medicines and vaccines
- Using tiny semiconductor chips as miniaturized platforms for rapid DNA and other biochemical analysis (*biochips*)
- Using artificially engineered materials for restoration, maintenance, or improvement of damaged tissue *or* even whole organs (*tissue engineering*)
- Moving from X-ray films to digital, highly detailed anatomic and functional scans of the body and interventional radiology that can replace surgery (*imaging*)

Converging technologies

Figure 4.1 illustrates how technological innovation beyond the traditional domain of healthcare shapes the biosciences.[2] Let's look at some of these diverse technologies in greater detail.

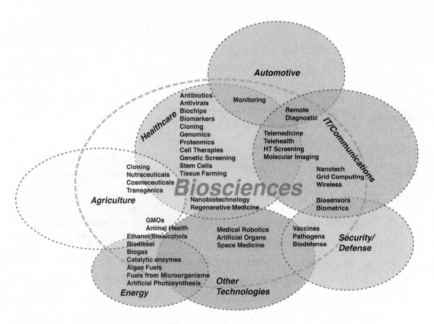

Figure 4.1 Bio-driven convergence

Telemedicine

Fast computers, abundant data storage capacity, and digital communications networks that can carry large amounts of data with high fidelity were originally developed for computing and data-intensive applications such as the Internet. However, these technologies are now used to transmit medical records and images from digital radiology and tissue pathology. Telemedicine enables remote assessment of patient data when the patient cannot visit a diagnostic facility nearby or when radiologists want to extend their expert reach. Sometimes more advanced or lower-cost expertise is located halfway around the world in countries such as India and Australia.

Remote diagnostics

Advances in wireless telemetry have increased patient mobility. Consider a patient with a serious heart condition who is able to transmit life-monitoring data. Data and alarms can now be wirelessly logged via implanted devices such as cardiac rhythm-management systems.[3] The information is then encrypted for privacy and security, and transmitted instantly to medical staff on call. Such remote diagnostics can prevent worsening heart conditions and even detect a malfunctioning device before it is too late. Similarly, the home or the car might become a valuable diagnostic environment for repeated observations about weight, cholesterol, or response to stress. For example, cars in the future may contain sensors such that a near-collision becomes an opportunity to assess your heart's functioning. Non-invasive measures will beam data about your blood pressure and heart rhythms via satellite to your doctor. Your health insurer will discount your premium if you participate in this monitoring program, because the near accident data will provide a more accurate stress test for your heart than the traditional tread mill.

Biological drug factories

Although plants and herbs have yielded natural medicines since ancient times, scientists are conducting extensive research on genetically engineered plants and microorganisms. DNA sequences from different sources are fused using recombinant DNA technology and then transferred to a new host organism (see Chapter 3, "Snapshot of the Biosciences"). Such genetic-engineering methods can produce drugs more efficiently or make them more potent. For example, researchers have developed a quicker, cheaper, and more efficient process to synthesize the antimalaria miracle drug *artemisinin*, which is extracted from the wormwood plant. By inserting the plant genes into bacteria, the microbes essentially become mini drug factories.[4]

Biochips

Conventional laboratory tests for the clinical diagnosis of disease can be time-consuming and often involve bulky equipment. By using micro-fabrication methods and large-scale electronic chip integration techniques developed for the semiconductor industry, it has become possible to apply several of these technologies to build *biochips*. Such chips were developed initially for DNA analysis but are now being extended to protein, antibody, and biochemical analyses (see Chapter 3 for a discussion of DNA chips).

Glucose sensors represent a major application of biochip technology. Biosensors couple molecules, such as enzymes and antibodies, to sensors (optical or electronic) that detect a reaction. For example, diabetics can determine blood glucose levels using small, lightweight glucose monitors and disposable, single-use biochips (strips). Glucose oxidase, an enzyme that binds to glucose, is printed on the chip. The patient puts a chip into the monitor, pricks her finger, and touches the chip. Glucose oxidase reacts with glucose in the blood, releasing a chemical. The monitor converts this chemical signal to an electronic one and displays the glucose concentration that is proportional to the electronic signal.

Tissue engineering

The availability of replacement parts for damaged skin, tissue, and even entire organs is the ultimate promise of a highly interdisciplinary field known as *tissue engineering*. This field combines innovations in materials compatible with the human body, molecular and cell biology to induce cell growth, mechanical aspects of artificial biological structures, and bioinformatics tools. The latter help to model and understand the genetic expression of the cells and help to

Biochips and individualized medicine

"Biochips will store every individual's genome. DNA and protein 'chips' will allow efficient diagnosis that could lead to very constructive preventative medicine. In this world, doctors will become more like shift workers, following individual-specific treatment regimens. And individualized medicine should bring doctors closer to patients through their more intimate knowledge of genetic predispositions and treatment impacts."

—*Dr. Christine Côté, Vice President, Emerging Technologies and New Ventures, Johnson & Johnson* manufacture, test, and evaluate engineered tissue. Tissue engineering uses living cells as engineering materials to synthesize organs and tissue, such as replacement cartilage in damaged or worn-out knees. The cells, which are extracted from blood and other fluid tissue, can be of many different types, depending on the function and final product desired. For example, *stem cells* might be used as primal building blocks to regenerate a variety of different structures in the body. This is also known as *regenerative medicine*.

Tissue and organs are typically constructed using synthetic scaffolds that are seeded with living cells. Using molecular self-assembly, the cells then grow on the scaffold into the final structure to be transplanted into a patient. These scaffolds can consist of resorbable materials, such as rat collagen, that will be fully resorbed in 4–38 weeks after the tissue has completely assimilated. They might also entail inert, nonresorbable materials, such as carbon nanotubes, which have high mechanical strength while being lightweight. Desirable scaffold properties include high porosity so that the living cells can grow into them, injectability into host tissue, and cost-effective manufacturing. Some scaffolds even serve as a means for drug delivery to a target tissue site after transplantation, such as in chemotherapy for cancerous tumors. A major advantage of tissue engineering is the use of your own cells, which makes an adverse immune response and rejection after transplantation less likely.

Imaging

We have come a long way since the hazy picture of traditional X-rays. New technologies permit a more precise view into the body, from computerized axial tomography (CT or CAT), magnetic resonance imaging (MRI), and positron emission tomography (PET) scans to ultrasound and MRI spectroscopy. These newer imaging techniques reflect the convergence of engineering, information technology, nuclear medicine, and molecular biology. Several important trends are reshaping the field of imaging.[5]

The first trend is greater anatomic clarity. *CT scanners* take many cross-sectional views of the body, similar to cutting thin slices right through the core. The separate images are then put together digitally into a complete three-dimensional picture. Images of the heart, heart valves, and coronary arteries can yield information on how the circulatory system is functioning. CT scans can also be used to find cancerous tumors throughout the body. They are increasingly used to perform "virtual colonoscopies," although the loathsome procedure of purging the colon is still required. In addition, we can now also buy a "total body scan." (The downside is false positives and associated unnecessary procedures.) MRI scans can also examine soft tissues, such as nerve tissue in the brain or spinal cord, by mapping the movement of hydrogen atoms within magnetic fields. MRI is also able to detect aneurysms in the arteries of the brain or abdomen, slipped discs in the spinal column, and tumors in soft tissue.

A second key trend is *molecular imaging*, which focuses on uncovering the disease process itself instead of just changes in anatomy. This requires injecting the patient with radioactive glucose that disseminates throughout the body, and then observing, via a PET scan, which cells are actively metabolizing the glucose. In addition to glucose, newer chemicals are on the horizon, including specific biomarkers of diseases that PET scans can trace. MRI can also be coupled with specific molecular agents to detect disease. For example, doctors can use iron oxide to detect small metastases of prostate cancer that other techniques would have missed. Anatomic imaging is fusing with molecular imaging to yield a more complete picture of the disease process. For example, PET-CT scanning of the heart using radioactive glucose not only provides a visual image of the heart, but can also see which parts are no longer functioning well.

A third major imaging development is *interventional radiology*, using imaging to treat specific diseases instead of to just diagnose them. Radiologists and surgeons can work together to fix aneurysms in the brain or abdomen, to repair internal bleeding, to insert stents, and to destroy tumors. For example, it is common now in colonoscopy exams to remove small polyps at the moment they are found.

Multidirectional synergies

An interesting corollary to this phenomenon of technological convergence is that the effect often is multidirectional. Just as the biosciences draw upon innovations created in fields such as engineering and agriculture, these other fields also draw upon and utilize technologies developed in the biosciences. We offer a few examples of such multidirected synergies to illustrate the full breadth of bioconvergence, which often extends beyond healthcare.

Biofuels

For some time, we have used biological materials such as plants and even animal waste for heating and cooking. During the 1970s, when gasoline shortages posed a threat to energy security, corn-based ethanol was developed as a substitute for gasoline. Ethanol is also promoted now as a "green" alternative. Other biofuels based on sugar cane, sugar beets, and non-food crop cellulosic plants such as switch grass are under commercial development. Cellulosic ethanol production is especially benefiting from biosciences' research into novel enzymes and ethanol-fermenting organisms, for more efficient bioethanol production.

Biodefense

Scientists are drawing on advances in biochemical sensors and microminiaturization from the biotechnology and semiconductor industries to detect pathogens and biological warfare agents. Early-warning biochips in which sensors detect chemical agents are being developed to defend against biological warfare, prevent bioterrorism, and control environmental quality. Because these compact systems are portable, they can also be more easily deployed in the field for rapid response to emergencies and sudden attacks.

Genetically modified plants

In agriculture, genetically modified crops have become more disease and pest resistant, thanks to recombinant DNA techniques. They include insect-resistant cotton and herbicide-tolerant soybeans. A well-known example is Roundup Ready soybeans, which were genetically altered to be more tolerant to Monsanto's broad-spectrum herbicide Roundup. This enables farmers to apply higher doses of pesticides, resulting in better crop growth and higher farm yield. However, consumer and government opposition to such genetically modified foods has been fierce for safety and political reasons.

Shifting industry boundaries

We have discussed how technologies can cross industry boundaries and cross-fertilize adjacent industries. Sometimes technological innovations flourish not in their native domains, but in another industry in which conditions might be more favorable. This is a phenomenon known as *technology speciation*, analogous to how natural species emerge and thrive as they adapt to new environments.[6] Favorable conditions can come in the form of abundant funding from investors, growing market demand, supportive government policy, or increased public awareness.

Figure 4.2 illustrates how individual technology markets, such as the telecom, defense, and energy markets, surge and become hot at different points in time. The surging market often grows by drawing upon latent innovations in neighboring industries, thus enabling these technologies to mature and flourish.

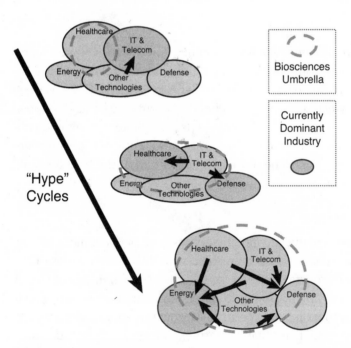

Figure 4.2 Evolution of hype cycle

Hot technology markets wax and wane. Recall the euphoria of biotechnology, the Internet boom, and the subsequent telecom revolution. When the telecom industry waned, the defense industry—and, in particular, the counterterrorism industry—picked up the slack following the September 11, 2001, terrorist attacks and adopted several "orphaned" telecom technologies. For example, high-power miniaturized lasers developed originally for fiber amplifiers in telecommunications systems are now being used in defense and counterterrorism applications such as range finding, laser radar (LIDAR), targeting, and biochemical agent detection. Another notable example is optical fibers that were originally developed for digital communications and later for sensing structural defects in bridges, automobiles, and aircraft. They are now finding uses in medicine as pressure sensors in catheters and as temperature sensors that are not prone to electromagnetic interference during MRI scans.

As these technologies cross industry boundaries, the boundaries themselves might shift or become blurred. Over time, the biosciences and adjacent industries benefit from a larger toolkit of innovations.

Healthcare and IT

The application of information technology (IT) to healthcare is some-times referred to as *healthcare information management, healthcare IT, medical informatics*, and *bioinformatics*. These terms are often used interchangeably and generally connote *medical information handling*. They use computers, software, digital data records, data storage, or communications networks in the delivery of healthcare. IT continues to have a tremendous impact on the full spectrum of healthcare, from preventing disease to diagnosing and treating it. Let's look at some current examples.

Medical information handling

The most significant use of IT in healthcare is electronic medical records (EMR) management, which includes recording, archiving, and reviewing electronic patient charts. This process includes not only computerized physician order entry systems (CPOE), but also computerized decision support systems (CDSS) to provide an audit trail for all medical decisions made and implemented. Although such systems promise efficiency and cost savings, implementation has been slow and expensive because of complex installation and testing of the software systems.[7] But several major initiatives are underway in areas such as the United Kingdom's National Health Service and, to a lesser extent, in the United States.[8] Electronic medical information handling also involves automation of care delivery as well as electronic systems for billing and scheduling.

Medical data storage

As more electronic data are created through medical image archiving, data storage requirements grow. Hospitals, doctors' offices, and other healthcare organizations that maintain electronic health records face the daunting task of managing data storage systems and the risk of malpractice litigation. Data storage costs have declined dramatically in recent years, but concerns remain about data protection after disasters such as fire, earthquakes, hurricanes (such as Katrina), or even a 9/11 type of terror attack. Disaster-recovery systems need to maintain privacy while building a scalable system that can absorb new data. Most hospitals outsource data storage to vendors who specialize in storage area networks. These vendors draw on expertise developed in other data-intensive domains, such as financial services, and now offer customized solutions for healthcare organizations.

Computers and drug discovery

Life sciences data have exploded in recent years, particularly in drug discovery. Calculations in life sciences, such as modeling of protein folding, elucidation of gene sequences in genome mapping, and simulation of biological systems, are computationally intensive. "Normal" computers don't offer sufficient computing power to tackle these problems. They require *supercomputers* that offer the highest processing capacity. The supercomputers are usually a cluster of many individual processing units functioning as one "computer." The individual units in this system often have a processing speed of regular computers, but in combination they are much faster. Currently, the fastest supercomputer, the IBM Roadrunner, runs about 50,000 times faster than a normal dual-processor PC.

Grid computing, an alternative approach to performing a large number of calculations, uses a network of remote computers instead of a single supercomputer. A central unit distributes work across these remote computers, called *clients*, which perform the calculations in parallel. The individual calculations are sent back to the central unit, which integrates all the data from individual clients. The largest distributed computing network, called Folding@home, simulates the protein-folding process. It uses the computing power of 3.3 million client computers, including personal PCs and even video game console systems. The combined computational power of the entire Folding@home network is greater than that of the largest single supercomputer in the world.

Remote monitoring and the wired home

Computing power has been doubling every two years during the last three decades. This is somewhat slower than the rate suggested by Moore's Law in 1965, which held that computing power would double every 18 months.[9] As the cost of computing power has declined, high-performance computing devices have become ubiquitous. A striking example is the wireless cellphone, which has driven more efficient chip designs to feed the tremendous demand for miniaturization. A significant new trend is *multicore* chip sets that use chip space more efficiently and require less power consumption. These more powerful processors were previously available only for high-end computing servers, but are now used in consumer electronics such as set-top boxes. This new computing power will fuel a new breed of implantable sensors that transmit large quantities of medical data about consumers in their natural setting. Examples include devices that monitor diabetic conditions in patients at home or heart rhythms in someone narrowly escaping an accident in a car (to see how well the heart functions under stress).

Reshaping bioinformatics

The conventional definition of bioinformatics is the application of information technology to the field of healthcare and biology. A broader definition is the access and use of any biological information in healthcare. Bioinformatics spans the gamut of genetic profiling, medical device monitoring, patient records tracking, and even recording of physiological parameters such as weight, skin

Convergence of biology and information

"The merger of biology and information is creating new innovations, such as tiny computers built into everything: eyes, ears, delivery devices, and microfeedback devices. Learning from biology is also being driven back into materials."

—*Dr. William Haseltine, founder and former CEO of Human Genome Sciences*

temperature, pulse, blood pressure, heart rate, and electrocardiogram signals. These layers of biological information are of use not only to doctors and nurses, but also to other "consumers" of patient-specific clinical information: pharmaceutical and medical device manufacturers (producers), Medicare and private insurance companies (payers), and group purchasing organizations (GPOs).

Figure 4.3 illustrates how a physician could use the full extent of a patient's information, from genetic to physiological data, to treat the patient holistically instead of just focusing on a particular disease. Biomarkers such as *tissue oxygen levels* can provide early warning in case the brain or other vital organs become oxygen starved. Patient-specific biological data ranges from biochemical information about drugs in the circulatory system to vital signs (such as blood pressure and temperature), to physiological data such as blood flow, tissue perfusion, and oxygenation. The power of this bioinformatics paradigm lies in the potential for a multilayered, integrated platform that provides critical real-time information to the physician, similar to the instruments panel in the airplane cockpit.

In addition to tracking illness-related changes in the body, doctors could track changes in the body induced by a patient's drug regimen *(pharmacodynamics)*. Such information is crucial to the pharmaceutical industry's efforts to improve data collection for post-market surveillance, patient by patient. This emerging paradigm of *personalized bioinformation* will not only improve clinical outcomes, but might also track the value that each stakeholder in the healthcare system— hospitals, physicians, payers, drug and medical devices manufacturers, healthcare IT service and equipment providers, sensor hardware

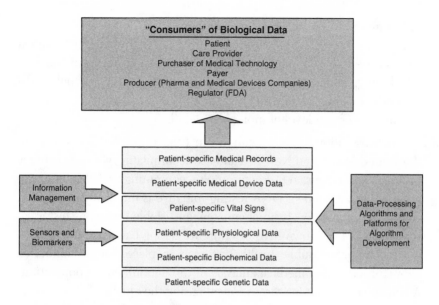

Figure 4.3 Bioinformation hierarchy

manufacturers, bioinformatics service providers, and the patient—contributes to the effective delivery of care.

Some of these new bioinformatics technologies likely will transform the spectrum of healthcare—prevention, diagnosis, and treatment—into a seamless continuum. For example, the same technology used to diagnose a disease might also be used to treat it. Ultrasound microbubble technology is an example in which tiny *gas bubbles* that are less than a thousandth of a millimeter are used to improve ultrasound imaging of diseased organs. Companies are also trying to deliver drugs to disease sites by encapsulating them in microbubbles, which doctors can then burst open using ultrasound waves to release the drug at the target site.

Let's further illustrate the great impact of bioinformatics by discussing gene banks, personalized medicine, systems biology, biosensors, and nano devices.

Bioinformatics and gene banks

When new methods for DNA sequencing became widely available, molecular sequence data grew exponentially. Since the sequencing of the first microbial genome in 1995, the genomes of more than 100 different organisms have been sequenced and large-scale genome sequencing projects have become routine. Researchers have built a strong repertoire of computational tools to translate accumulated data into biologically meaningful information. Using these "virtual workbenches," scientists can study both healthy and diseased organisms at an unprecedented level of molecular detail.

Various national DNA databases in France, the United Kingdom, the United States, and New Zealand are accumulating genomic profiles of large populations, requiring complex computing and database systems. The first national DNA testing initiative is underway in Iceland, where the population is homogeneous and stable, excellent historical records exist from town halls or churches dating back centuries, and a well-documented public health system is in place. These conditions allow for a detailed medical analysis of the nation's entire population. A similar study is under way in Canada using the Quebec founder population, which started with just 2,600 immigrants between 1608 and 1760. It has since grown 800-fold with little marriage outside the group.

Personalized medicine

Gene expression research suggests that effective clinical therapy needs to better account for the individual's unique response to therapy.[10] The trend toward personalized medicine therefore addresses not only the targeted delivery of the drug to a disease site (targeted therapeutics), but also the identification and use of therapeutic indicators, such as genetic, biochemical, and physiological markers, to customize the best therapy for each patient. Given the variability in patients' responses to treatments, an urgent need exists for pre- and post-market surveillance of adverse reactions to drugs and implanted devices. Companies such as Theranos are developing biosensors in the form of portable blood assay kits to assess adverse drug reactions in patient populations. These kits use micro fluidics-based assay systems to detect changes in biochemical markers directly induced by the drug, and then wirelessly relay the results to medical personnel.

Systems biology and holistic medicine

Physicians often remark that the healthcare system treats diseases and not the patient. Considering the wide range of diseases that senior citizens are being treated for simultaneously, complications from

Genomic databases

The first genomic databases, such as GenBank, European Molecular Biology Laboratory (EMBL) Nucleotide Sequence Database, and DNA Bank of Japan (DDBJ), used novel computational methods for sequence similarity searches, functional analysis, and structural predictions. One of the first breakthroughs in the area of bioinformatics was the rapid sequence database search tool Basic Local Alignment Search Tool (BLAST), which has become an indispensable tool in the everyday life of biomedical research. BLAST compares nucleotide or protein sequences and calculates the statistical significance of matches.

Automatic sequencing was the forerunner of powerful high-throughput screening of various kinds of biological data, such as single-nucleotide polymorphisms (SNPs) and expressed sequence tags (ESTs). Subsequently, other novel high-throughput methods, such as serial analysis of gene expression (SAGE) and DNA microarrays, have been developed to analyze the transcriptional program of a cell, tissue, or organism at a genomic scale.

All these novel experimental procedures are associated with information technology in a symbiotic relationship. The use of high-throughput experimental procedures in combination with computational analysis has revealed a wealth of information about important biological mechanisms. Past and future discoveries are transforming database technologies and computational methods to facilitate the integration and visualization of these different data types, ranging from genomic data to biological pathways.

the interactions of these multiple therapies can be life-threatening. Therefore, a critical need exists to treat the patient as a whole person. This can be done by having interconnected sensors, which are commonly used in cars, aircraft, and weather forecasting, to monitor changes system wide.[11] Doing this in humans requires implantable biomedical *smart sensors*.[12] Such systems could greatly enhance the quality of the information that is gathered, for example from multiple organ and tissue sites in the body, and would represent a major departure from the present localized, or focal, sensing strategies. A *systems biology* perspective is already emerging in the management of diseases such as cancer. Genetic markers, CT scan images of the patient's anatomy, and changes in blood flow to tumor sites *(angiogenesis)* detected via noninvasive infrared sensors are being combined to fight cancer.[13]

Biosensors

Biosensors are emerging as an important adjunct to patient care. Such sensor platforms include drug delivery systems in which sensors regulate drug dosing *(integrated therapeutics)*. Traditional pharmaceutical companies, many of which historically shed their capabilities in device technology, are now entering into development collaborations with medical device manufacturers. Examples include Stryker and Zimmer's orthopedic components with antibacterial coatings; Boston Scientific, Johnson & Johnson, and Medtronic's drug-eluting stents; and Medtronic's proposed integrated glucose monitoring and insulin pump system for diabetes care. These developments prompted the FDA to create an office to handle combination devices in 2005. The biosensors market is attracting the attention of nonhealthcare companies as well, such as IBM, Intel, and Cisco. Even Google has declared its interest in organizing and monetizing biological and medical information through selected investments in bioinformatics start-ups.[14]

Nano devices

Nanotechnology is defined as the creation and manipulation of artificial devices operating at a scale of a billionth of a meter. Nanoparticles offer unique physical and material properties not available at the macro scale. Examples of nanoscale miniaturization include carbon nanotube sensors for extremely sensitive CO_2, virus, or DNA detection (see Nanomix, Inc.). In 2000, the National Science Foundation and other federal agencies in the United States designated nanotechnology as a major national research priority, given its potential to advance science and technology in virtually all industries, including biomedicine. Many other countries, such as Australia, Canada, Japan, China, Korea, and Singapore, have followed suit. The emerging field of *nanobiotechnology*, which combines healthcare, biomechanics, and IT, could be a critical enabler in the future. Nanotechnologies could help improve drug delivery and extend the life of currently available drugs. These emerging technologies offer new approaches to understanding complex biological systems, measuring drug effects, predicting individual outcomes, and improving the R&D and manufacturing processes.

Nanoscale sensors can be inhaled, infused through the skin, or even ingested. In the latter case, the sensors will likely attach to binding agents that target specific receptor sites in diseased tissues of the body. Some companies, such as Nanospectra Biosciences, Inc., are developing cancer "targets" based on nanoscale structures to accurately destroy tumors with focused heat or light. Others, such as Nanoscan Imaging Inc., are developing nanoparticles as high-contrast agents for CT (X-ray) scans to better diagnose heart disease and cancer at early stages without the toxic side effects to the liver from current diagnostic contrast agents such as iodine. However, concerns still remain about how nanoparticles can be safely flushed from the body after therapy. For example, even if nanomarkers are proven to be relatively inert or benign in the body, consumers' fears might prompt an unwarranted backlash, similar to the concerns expressed about genetically modified foods.

Commercialization challenges

Although the biosciences sector might foster rich and unexpected technology crossover from other industries, the resulting products and services still have to meet strict market and regulatory standards. The pace and direction of technological developments are often hard to predict because they can encounter unexpected obstacles. Let's look at some of the key challenges facing the commercialization of bioscience technologies.

The power of biomarkers

"In the past, you'd ask the patient how they feel and have a scale. Biomarkers are more objective, being able to indicate if your drug is hitting its target, affecting the target, and affecting the state of the actual disease."

—David Lester, PhD, Clinical Technologies, Pfizer

Technological challenges: biocompatibility

With improvements in mini-batteries and other devices (such as faster, cheaper, smaller, and wireless), new bioscience innovations will find commercial application. However, various technological challenges remain that can thwart commercialization. For example, the body might adversely react to biosensors or other implants. They need to be designed to avoid an undesirable immune response while the devices perform their function. Although *biocompatible* materials are not yet widely available, researchers have made progress in engineering bio-materials using natural compounds, a field known as *biomimetics*. Examples of biomimetic applications include recombinant protein-based drug coatings on tissue scaffolds to stimulate tissue healing and regeneration (for example, the work of BioMimetic Therapeutics, Inc.).

Regulatory challenges: clinical safety and efficacy

Clinical safety and efficacy is necessary for successful adoption of any bioscience technology that directly involves humans. For radically new products in which several technologies converge, such as nanosensors or hybrid drug-device combinations, testing to high drug standards might become prohibitively expensive. Companies might therefore choose *not* to pursue such risky technologies, knowing that they won't be able to recover their commercialization costs. The funding, training, and quality control requirement of the FDA approval process might discourage the commercialization of novel, high-risk technologies.[15]

Political challenges: privacy and ownership

The Health Insurance Portability and Accountability Act (HIPAA) of 1996 is the key legislative framework that now guides the protection of private patient medical information in the United States.[16] This leg-islation is particularly relevant to the commercialization of implantable biosensors. The mining of biological data from patients will likely create private patient data that could be compromised. HIPAA regulations provide the patient with rights of access to, owner-ship of, and authorized release of personal medical information. Data security poses a significant challenge because it might be difficult to defend against cyber attacks on healthcare IT systems, even if the law mandates the security and privacy of patient data.

Also, it's not always clear to what extent medical data is "owned" by the patient or others. Consider the simple case of a blood test ordered by a life insurance company to issue a policy. Who owns the information resulting from the blood test? Is it the patient who supplied the blood, the life insurance company that paid for the test, or the doctor or hospital that interprets the results? The control of information includes not just the ability to access, create, modify, package, or derive benefit from the data, but also the right to assign these privileges to others.[17]

Ownership or access becomes even more complicated if a person's private medical data, when combined with that of others, can serve a significance public interest to stem an epidemic or advance research. An example is the data generated from the Human Genome Project, in which the legal status of genetic material and genetic information is unclear. Even though many hospitals consider the records in their systems to be their property, many patients argue that their medical information is really theirs.[18] Information ownership might need to be ceded to the hospitals and third parties on a "need-to-know" basis, but unrestricted patient access must be allowed for their own data.[19]

Risk of lawsuits can further thwart the adoption of innovative technologies. The rise of healthcare costs in the United States has been attributed in part to escalating medical malpractice awards in jury verdicts. As U.S. malpractice costs and insurance premiums escalate, medical device and drug companies might become increasingly risk averse, sticking with accepted technology and treatment methods.[20] Although this might be the policy goal of the malpractice regime, it also means that the development of technologies with great clinical promise might be delayed.

Social challenges: consumer advocacy

If people don't fully understand a new technology's impact on health, safety, or privacy, they will support consumer advocacy groups and watch dogs. For example, implantable biosensors raise concerns about safety, abuse of medical data, and loss of control, so advocacy groups are being formed. Law firms will also become keenly interested in case the new technology is unsafe by causing injury or death. A current example is the poorly understood risks of residual nanoparticles in the body, which might prompt class-action lawsuits in the future.

Grass-roots advocacy organizations such as Health Care for All (HCA), National Alliance on Mental Illness (NAMI), and Consumers Union are becoming more visible and vocal.[21] They support educational programs and push for action on key social and health issues with donor support. For example, the consumer advocacy group Public Citizen lobbied the Centers for Medicare & Medicaid Services (CMS) to limit or deny reimbursement for a new, and purportedly unproven, device for electrical stimulation of the *vagus* nerve to treat depression.[22]

The road ahead

It's unknown how the previous challenges will affect biomedical progress. Technological failures and setbacks will occur. We know from history that many important technologies have taken decades or even centuries to achieve full market acceptance.[23] For example, the first patent for the fax machine was filed more than 160 years ago in 1843, but the technology became a commercial success only in the 1980s. The Internet took multiple decades to move from the military and academic communities to private homes and businesses. In the early days of cardiac and transplant surgery, most patients died from complications before heart bypass and transplant tissue typing were perfected. More work is needed to achieve the full promise of the biosciences. Chapter 5, "Business of Biomedicine," reviews the main healthcare industries that play key roles in bringing biomedical technologies to the market. Chapter 6, "Healthcare Under Stress," examines the conditions of the broader healthcare system that are needed to finance, regulate, and deliver biomedicine.

Endnotes

[1]This chapter was written jointly with Nanda Ramanujam, Ph.D., founder of the Ramanujam Consulting Group, an independent management consultancy focused on growth strategies for early-stage technology businesses. He has also been a founder or key member of several startups spanning the telecom, medical devices, and solar power industries.

[2]This is an expanded and modified chart originally developed by Scott Snyder, Ph.D., of Decision Strategies International, Inc. (www.thinkdsi.com).

[3]"Patients with Defibrillators Take Wireless Technology to Heart, Home," *Associated Press* (8 August 2006).

[4]J. D. Keasling, et al., "Cheap, Simple Microbial Factories for Antimalarial Drug,"

Nature Biotechnology (July 2008).

[5]Stephen C. Schimpff, *The Future of Medicine: Megatrends in Health Care That Will Improve Your Quality of Life* (Nashville, Tenn.: Thomas Nelson, Inc., 2009).

[6]Ron Adner and Daniel A. Levinthal, "Technology Speciation and the Path of Emerging Technologies," chapter 3 in *Wharton on Managing Emerging Technologies* (New York: Wiley, 2000).

[7]"Electronic Health Records in Ambulatory Care—A National Survey of Physicians," *New England Journal of Medicine* 359:50, no. 1 (3 July 2008): 60.

[8]J. C. Goldsmith, "The Healthcare Information Technology Sector," in *The Business of Healthcare Innovation* (New York: Cambridge University Press, 2005).

[9]A law, commonly attributed to Gordon Moore, legendary founder of Intel Corporation, postulated that transistor count per microprocessor—and, therefore, computing power—is likely to double every 18 months.

[10]D. Christensen, "Targeted Therapies: Will Gene Screens Usher In Personalized Medicine?" *Science News Online* 162, no. 11 (14 September 2002): 171.

[11]J. M. Kahn, et al., "Mobile Networking for Smart Dust," in ACM/IEEE International Conference on Mobile Computing and Networking (MOBICOMM 99), Seattle, Wash., 17–19 August 1999. D. Steere, et al., "Research Challenges in Environmental Observation and Forecasting Systems," *Proceedings of MOBICOMM* (2000): 292–299.

[12]L. Schwiebert, et al., "Research Challenges in Wireless Networks of Biomedical Sensors," *Proceedings of the 7th Annual MOBICOMM* (2001): 151–165.

[13]J. O'Shaughnessy, "Molecular Signatures Predict Outcomes of Breast Cancer," *New England Journal of Medicine* 355:6 (10 August 2006): 615–617.

[14]*Continua Health Alliance*, an Intel-led consortium of more than 20 companies, focused on standardizing the management and interoperability of medical devices. Consortium companies include Cisco, GE Healthcare, IBM, Intel, Medtronic, Motorola, Panasonic, Philips, Samsung, and Sharp.

[15]Recommendations of the Secretary of the Department of Health and Human Services in "Report to Congress on the Timeliness and Effectiveness of Premarket Reviews" (August 2003).

[16]Public Law 104-191, Health Insurance Portability and Accountability Act (21 August 1996). Available at http://aspe.hhs.gov/admnsimp/pl104191.htm.

[17]D. Loshin, "Knowledge Integrity: Data Ownership" (19 July 2002). Available at www.datawarehouse.com/article/?articleid=3052.

[18]G. J. Annas, "A National Bill of Patients' Rights," *New England Journal of Medicine* 338 (1998): 695–699.

[19]R. Schoenberg and C. Safran, "How to Use an Internet-Based Medical Records Repository and Retain Patient Confidentiality," Center for Clinical Computing, Beth Israel Deaconess Medical Center, Harvard Medical School. Available online at www.

hipaadvisory.com/action/patientconf.htm.

[20]S. Seabury, "Does Liability for Medical Malpractice Drive Healthcare Costs and Technology Adoption?" American Society of Health Economists (ASHE) inaugural conference on Economics of Population Health (6 June 2006).

[21]Health Care for All societies in various states (www.healthcareforall.org, www.njshca.org, www.hcfama.org, and www.vthca.org), National Alliance on Mental Illness (www.nami.org), and Consumers Union (www.consumersunion.org).

[22]B. J. Feder, "Battle Lines in Treating Depression," New York Times (10 September 2006).

[23]B. Chakravorti, The Slow Pace of Fast Change: Bringing Innovations to Market in an Interconnected World (Boston: Harvard Business School Publishing, 2003).

5

The business of biomedicine

with Jim Austin[1]

This chapter examines five important segments of the U.S. health system: pharmaceuticals, biotechnology, medical devices, medical diagnostics, and prevention/disease management (see Figure 5.1). While the first four are established segments, the latter—prevention and disease management—may rapidly evolve in the future. The dynamics of each segment can be a barrier to or a supporter of new technologies because they constitute the "playing field" on which future biomedical treatments will be developed. Our discussion focuses mostly on the United States because progress in the biosciences greatly depends on U.S. reimbursement, innovation, and regulatory conditions.

The pharmaceutical industry

This industry is clearly under pressure. Major new drug releases are declining and public trust has diminished due to recalls and lawsuits, while aging populations are exerting political influence to lower prices. The call for basic universal coverage and greater use of generics pose further problems. In the developing world, where enormous demand exists, few can afford patented drugs. Not surprisingly, the market values of pharmaceutical companies have declined.

Drug discovery has always been a very high-risk endeavor, involving multiple stages. Creating a new drug from discovery to market typically takes more than 12 years, at an estimated cost of more than $1 billion (up from $54 million in the late 1970s). Yet as Table 5.1

High failure rate

"Only 3 out of 10 drugs (that make it to market) typically recoup their capital costs."

David Block, CEO, Rexton Pharmaceuticals, former Chief Operating Officer of Celera

shows, the success rate is not high. For every 5,000 compounds evaluated, 1 makes it to market. When patents expire and generic substitutes enter the market, drug prices typically drop 70% to 80%.

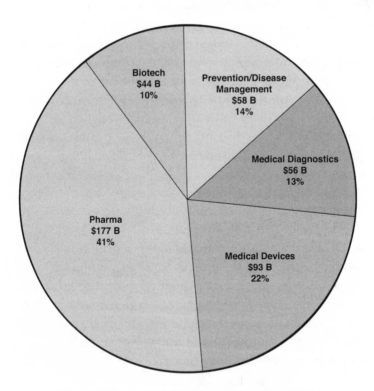

Source: Frost & Sullivan, CDC, Thomson Healthcare, IBIS, Data Monitor, and World Research Group, IMS Health, MIDAS, MAT Dec 2007

Figure 5.1   2007 U.S. revenues (billions)

Table 5.1 U.S. drug-development phases

Phase	Years	Number of Compounds
Preclinical	3.5	5,000
Promising new compounds first undergo preclinical testing in animals. If the preclinical data is positive, the U.S. Food and Drug Administration (FDA) designates the compound as an investigational new drug (IND). Research then moves to clinical testing in people through clinical trials in phases I, II, and III.		
Phase I	1	5 enter trials
Is it safe? This phase determines how the drug works in healthy participants enrolled in a trial. Researchers examine the mode of action (how the drug exerts its effects), safety, and side effects. The overall efficacy of the medication in patients is not established in this phase, just its safety.		
Phase II	2	
Is it effective? This phase determines whether a drug's clinical activity might be beneficial against a particular disease or condition. A drug reaches Phase II only after the FDA has reviewed the Phase I data and concluded that the drug is safe enough to test in patients. At this point, a larger group of patients (such as more than 100) is enrolled, and rating scales specific to a condition or disease are used to record data.		
Phase III	3	
Can it be used in more patients? At this point, the medication is ready to be studied in a larger population, such as 1,000 patients, with even more advanced rating scales and clinical measures. In recent years, the trend has been to include "real-world" measurements, such as how patients' daily living activities are improving.		
FDA review	2	1 passed
The company files its New Drug Application (NDA) with the FDA for its review and approval, based on the clinical data supplied from the Phase I–III clinical trials.		

Table 5.1   U.S. drug-development phases

Phase	Years	Number of Compounds
Phase IV		

Can it be used for different medical conditions or sub-groups of patients? At this phase, the FDA has already granted approval, but the study might identify an additional use or gather more safety information from a larger group of patients. These studies can also provide information on how the drug might be best used or best combined with other treatments. Sometimes Phase IV studies establish effectiveness in a subgroup of patients, such as children or patients older than age 65.

Source: National Psoriasis Organization, www.psoriasis.org/research/pipeline.php

Diminishing drug pipelines

Several signs show that the flow of new drug products from R&D investment (drug pipelines) is dwindling. Total blockbuster sales—defined as products that yield more than $1 billion in revenues each year—are projected to increase at a compound average growth rate of only 1%, compared to approximately 9% percent during the period 2002–2005.[2]

The approval of entirely new compounds that have never been submitted to the FDA, known as new chemical entities (NCEs), fell from a high of 53 in 1996 to just 17 in 2007—the lowest recorded number since 1983.[3] Figure 5.2 shows how pharmaceutical companies raised spending levels on new drug pipelines, but with diminishing results: Annual R&D spending doubled during the 1990s, while the number of new chemical entities declined.

One strategic response of pharmaceutical companies has been to "extend" the patent protection by introducing minor variations to the initial product. For example, Abbott's drug Depakote, a treatment for bipolar disease and epilepsy, was reformulated to Depakote ER (extended release). The new formulation reduced the number of daily pills required, to improve patient compliance, gain incremental patent coverage, and sustain competitive differentiation from alternatives. However, reimbursement agencies and insurers are increasingly penalizing such "line extensions" by requiring patients to pay a

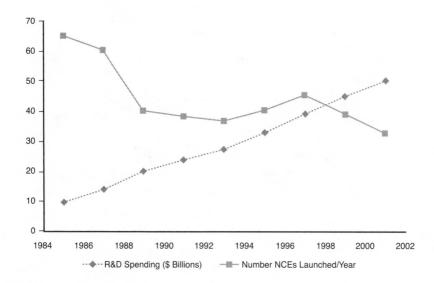

Source: IMS Health, Pricewaterhouse Coopers

Figure 5.2 The innovation challenge

Past performance versus future prospects

"In the next five years, the pharmaceutical industry will need to bring to market four times the number of products they brought to market in the past five years, to maintain a comparable level of revenue performance. This is highly unlikely, given market forces, competitive pricing pressures, and the number of scientists, investigators, and sales/marketing professionals needed to make this happen. Growth expectations as well as the need to survive will necessitate significant innovation and new models in all aspects of the drug life cycle."

Terry Hisey, U.S. Managing Principal, Life Sciences and Healthcare, Deloitte & Touche, LLP

premium (through higher copayments) for the reformulated but medically equivalent old drug.

Several explanations have been suggested for the decline in the number of breakthrough products from pharmaceutical companies, from a streak of bad luck to much of the "low-hanging fruit" having been harvested. Akin to the oil or gas industry, a limited number of "elephant fields" might exist for scientists to discover. We might be running out of naturally occurring small molecules that possess truly remarkable medicinal properties. However, unlike oil or gas deposits, the pharmaceutical industry can create new compounds via "rational drug design" and deploy new genetic-screening technologies (as discussed later). For example, the entire field of drug development was focused on approximately 500 basic cellular or physiologic mechanisms during the past 50 years, but the potential universe of targets is estimated to be many times larger.[4] But as Jean-Pierre Garnier, the former CEO of GlaxoSmithKline, argues, the organization of R&D in large pharmaceutical companies might be literally "too big," requiring revamped, smaller, more entrepreneurial groups focused on specific diseases.[5]

Large pharmaceutical companies also traditionally concentrated on the wealthier countries, especially the United States, Europe, and Japan, which together represent more than 80% of total drug sales and profits. The needs of billions of people in the developing world constitute enormous unmet demand, but with few resources to spend. The CEO of GlaxoSmithKline, Andrew Witty, recently announced a major push into developing markets and concluded a deal with a company in South Africa to market "branded" generic products. Such behavior is unusual for large pharmaceutical companies, as is Witty's decision to invite government reimbursement agencies early on to help evaluate the promise of his company's new drugs.

Considering these numerous challenges, large pharmaceutical companies are pursuing a number of interrelated strategies:

- More creative alliances with smaller biotechnology companies to fill the diminishing drug-development pipeline of larger companies
- Greater focus on global market opportunities, often utilizing joint ventures with local entities to lift sales
- Mergers and acquisitions to consolidate strong firms and gain greater scale, with more spending on R&D and less on marketing and sales
- Smaller, more entrepreneurial R&D groups with tighter efficiency controls, including outsourcing high-cost efforts

overseas (such as Phase II and Phase III trials often requiring thousands of patients)

- A move away from a mass-oriented blockbuster model and me-too drugs to a broad portfolio of niche or smaller patient–group drugs
- Changing the business model to better fit "personalized medicine" and disease "niches," as enabled by new discovery techniques

How well these and other strategies will work is still unclear. Looking forward, the traditional pharmaceutical companies might not retain enough power to play the lead role in the commercialization of promising new bioscience technologies.

An easy target

"It seems unlikely that the pharmaceutical industry will survive in its current form: The industry is rapidly becoming the tobacco industry of this decade, becoming too easy a target."

George Milne, Executive Vice President for Global R&D (retired), Pfizer, Inc.

The biotechnology sector

The scope of biotech treatments is quite broad ranging from cancer and diabetes drugs to growth hormones and vaccines. Early biotechnology companies were mostly small, start-up R&D entities seeking to develop novel therapeutics based on recombinant DNA (rDNA) technology (as discussed in Chapter 2, "A Short History of the Biosciences"). Today the sector typically includes entrepreneurial companies using new or innovative R&D technologies to develop novel drugs, diagnostics, or research tools, as well as large pharmaceutical firms. With more than 1,200 private and 300 publicly traded biotech firms, the sector has produced significant advances in treating cancer (Avastin, by Genentech), kidney failure (EPO, by Amgen), multiple sclerosis (Avonex, by Biogen Idec), respiratory syncytial virus (RSV)[6] (Synagis, by MedImmune), and various diagnostic, genetic profiling, and animal cloning capabilities.

Pharmaceutical companies during the past century focused their R&D on massive screening for promising chemical compounds and synthetic chemistry to explore and improve their candidate drugs. The biotechnology companies operate from a new paradigm: gene-based investigations, seeking to create drugs by recombinant and other biological processes. Initially, biotech companies focused on proteins with known physiological functions. Scaling up production of these proteins spawned advances in manufacturing processes, greatly increasing production via giant "bioreactors." Some of these products, such as insulin for diabetics and clotting factor for hemophiliacs, were already available to patients but relied on low-yield, more expensive natural sources.

The need to search more broadly

"Today's drug therapy is based on the manipulation of approximately 500 molecular targets. If one assumes that each multifactorial disease (such as diabetes, cancer, or spina bifida) may depend on contributions from 5 to 10 genes and accepts the estimate that about 100 to 150 diseases are based on altered gene functions, one could expect 1,000 'disease' genes. Each of these genes or gene products should be amenable to functional modification by at least three to ten additional proteins. Therefore, the number of drug targets to be discovered and utilized appears to be in the range of 3,000 to 10,000, significantly greater than 500. The initial approach to identify and utilize these targets by brute force has largely failed. In spite of the massive use of high-throughput screening (HTS), combinatorial chemistry, and genomic data, the productivity of drug research has not significantly improved."

Jürgen Drews, "Changing Scientific Patterns in the Quest for New Drugs," 4th Annual Conference on Molecular Structural Biology, September 2001

Overall, biotech's historical financial results have been disappointing. Excluding the results from the five most successful companies (which include Amgen and Genentech), profitability and stock market returns for the remaining biotech firms have been poor.[7] So how has the business been able to survive? First, the industry has

generated cash from multiple sources: venture capital (VC), government grants, and public equity. In addition, established pharmaceutical companies have created extensive alliance and ownership
arrangements with many biotechnology companies. In the past, pharmaceutical companies were vertically integrated enterprises, typically
forming external alliances only to fund specific research with universities or for access to markets via foreign pharmaceutical companies.
But times have changed for most pharmaceutical firms.

In 1978, Genentech reached an agreement with Eli Lilly to produce recombinant insulin in return for global manufacturing and marketing rights. This was the first time an established pharmaceutical
company formed an alliance with another for-profit entity to run a
proprietary R&D program. At the time, pharmaceutical companies
believed that biotechnology firms offered interesting new technologies for improving manufacturing, and possibly drug discovery, but
that they lacked the experience and capabilities to be a full-fledged
R&D partner. Only in the 1980s did biotech firms begin to change
these perceptions. In addition to obtaining more short-term financing,
they gained improved royalty-sharing agreements for future products.

Today such collaborative efforts have greatly expanded (see
Figure 5.3). Several hundred new strategic alliances are completed
every year, with more than 2,000 already in existence. How important

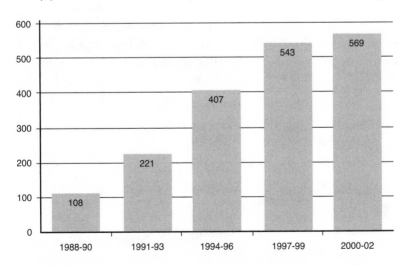

Source: Gary P. Pisano, Science Business, Harvard Business School Press, 2006, pg. 104

Figure 5.3 Total R&D alliances for the top 20 pharmaceutical companies

are such alliances to pharmaceutical companies? Industry experts estimate that more than 65% of the clinical pipeline of all pharmaceutical companies reside within the biotechnology sector, and that at least 50% of these are novel therapeutics with substantial new market potential.[8] Again, Genentech seems to have led the way. In 1990, Roche purchased 60% of Genentech for $2.1 billion, believing that Genentech could deliver a string of new therapeutic products over many years. Roche bought most of the remaining shares in 2009. Until recently, Roche managed this relationship as a separate research and development group to maintain Genentech's culture, capabilities, and output. However, R&D productivity continues to be a big problem, for both established pharmaceutical companies and many new biotechnology firms.[9]

As a result of these forces, large pharmaceutical companies play an ever increasing role in biotechnology. Through organic development, alliances as well as acquisitions of smaller biotech firms, they now dominate the top ten list of global biotech companies. In order of global revenue, the top ten are Amgen, Genetech/Roche, J&J, Novo Nordisk, Lilly, Sanofi-Aventis, Abbott, Merck, Schering Plough, and Wyeth. Amgen is the largest biotech firm with $16 billion in global revenue, whereas Wyeth is the smallest of the top ten, with $2.2 billion derived from biotech sales. Amgen and Genentech (now owned by Roche) are the only two in the top ten that were originally started as biotech ventures, with Amgen remaining as the sole independent one. In 2007, there were 106 blockbuster drugs in biotech (representing sales of at least $1 billion worldwide each), compared to only 36 biotech blockbusters in 2000. The top selling biotech drugs in 2007 were

- **Enbrel** (by Amgen/Wyeth for rheumatoid arthritis)
- **Aranesp** (by Amgen for anemia)
- **Remicade** (by J&J/Schering Plough for rheumatoid arthritis)
- **Rituxan** (by Genentech/Biogen Idec for non-Hodgkin's lymphoma)
- **Neulasta** (by Amgen for neutropenia)
- **Procrit** (by J&J for anemia)
- **Herceptin** (by Roche for breast cancer)
- **Epogen** (by Amgen for anemia)

A striking feature is how dominant traditional pharmaceutical firms have become in biotech and how commonly alliances are formed across and within these two industries to develop, manufacture, market, and sell biotech drugs.

Medical device industry

Medical devices represent another key sector for business opportunities in biomedicine. They range from highly sophisticated electronic implants to simple surgical staples, spanning nearly every medical specialty. About 55% of the industry is equipment and the remainder consists of supplies.[10] Within the United States, the medical equipment and supplies industry exceeds $100 billion in annual revenues and is primarily split between three segments: cardiovascular, orthopedics, and neurology (see accompanying sidebar).

All three segments are highly dependent upon continuing product innovation, which historically has lowered the total cost of treatment while improving outcomes. The major manufacturers in all three segments have strong, close relationships with the physicians who use (and often help improve) their products. The definition of a medical device (based primarily on electromechanical principles) versus a pharmaceutical (based on chemistry, biology, and, increasingly, genetics) is blurring. For example, such hybrids as drug-eluting stents (DES) are metal wire "tubes" coated with a drug to support collapsing arteries. In addition, pricing control for medical devices is beginning to shift from manufacturers and physicians to hospitals and payers. This is most apparent in orthopedics, where group purchasing organizations (GPOs) and hospitals are increasingly offering only a few brands and products, limiting the orthopedic surgeon's previously unfettered choices. Hospitals and GPOs are extracting "volume discounts" from the manufacturers, effectively lowering the price of different products.

Cardiovascular device segment

Cardiovascular devices cover coronary (heart) stents, pacemakers, heart valves, and other implantable heart monitors. This is the largest segment of the medical device sector and has grown at more than 12% per year during the past decade. Growth is expected to continue

Major device segments and players

The medical device industry typically is divided among the following three segments, with the major companies listed as well:

- **Cardiology**—Major product areas include cardiac rhythm management (such as pacemakers and implantable defibrillators) and interventional cardiology (stents/angioplasty). Leading companies include Boston Scientific, Johnson & Johnson, Medtronic, St. Jude Medical, C. R. Bard, and Edwards Lifesciences.

- **Orthopedics**—Major subsegments include spinal, hip, and knee implants, among other devices. The top six companies—Johnson & Johnson, Stryker, Zimmer, Smith & Nephew, Medtronic, and Biomet—control more than 70% of the market.

- **Neurology**—This emerging, fast-growing segment includes deep brain stimulators and drug pump technologies to treat chronic pain, motion, and neurological disorders, potentially including epilepsy and depression. Major companies include Medtronic, St. Jude Medical (ANS), Boston Scientific (Advanced Bionics), and Cyberonics, which together represent more than 80% of the market.

at more than 10% due to technological advances, the rising incidence of heart disease in our aging population, and the near-epidemic of obesity that can increase the incidence of type II diabetes and, in turn, heart failure.

Cardiac rhythm management (CRM) devices, including defibrillators and pacemakers, represent more than 50% of the cardiovascular device segment. However, in the last two years, the number of patients receiving defibrillators in the United States has actually declined for the first time since they were introduced in 1985. The problem is that no simple test exists to determine who should receive these expensive devices.[11] With recent safety concerns—in June 2005, Guidant Corporation recalled 29,000 implanted defibrillators due to electrical failures—total U.S. sales are flat or down.[12] The second-largest cardiac segment is coronary stents. Since their introduction in

2002, drug-eluting stents (DES) have dominated the product group and provided extraordinary growth.

Orthopedic device segment

The orthopedics segment consists of spinal, knee, hip, shoulder, and trauma implant products (such as plates and screws to repair broken bones after a car crash). Until recently, this segment has grown at nearly 13% per year, driven by new technologies such as partial hip replacements. This enables less invasive, more flexible surgeries using lighter and stronger parts, as well as new materials. The latter include demineralized bone, faster-setting biodegradable bone "putty," carbon fiber–reinforced polymers, and ceramic part construction (to replace metal). More fundamentally, this segment is experiencing a transition from mechanical to biological innovations and hybrids of the two, with newer technologies such as hip resurfacing and bio-absorbable materials. Manufacturers are developing "smart devices" that feed back data to the healthcare professional about relative position, wear, and estimated replacement life of the implant. Major suppliers in this segment are combining advanced informatics capabilities with less costly foreign-made implants to prolong product performance and to reduce cost.[13]

However, the orthopedics market momentum has recently slowed as GPOs and cost-containment measures have put pressure on manufacturers to hold or lower prices. In addition, congressional committees and the Justice Department are challenging the previously unfettered relationship between manufacturers and orthopedic surgeons in developing products, to protect against undue conflicts of interest.

Neurological device segment

Products in this segment send electrical signals to various parts of the body to manage pain, seizures, and other nerve-related problems. This segment is relatively immature, with less than $1 billion in total sales. But it might evolve rapidly due to clinical evidence showing its superiority—in terms of cost or efficacy—relative to alternative drug or surgical treatments. Two areas, in particular, hold promise for enormous growth: carotid stents to reduce strokes and implantable neurostimulation devices for pain management. Neurostimulation devices send electrical charges to different nerves or areas of the body to interrupt the flow of pain signals to the brain. Medtronic, St. Jude Medical (which acquired Advanced Neuromodulation Systems), Boston Scientific, and Cyberonics are currently the leaders in this area. The demand in the United States for approved devices, plus promising ones in clinical trials, is estimated to entail 76 million patients, yielding a potential market value of $35 billion.[14]

Even though the aggregate medical devices industry—cardiovascular, orthopedics, and neurology—is expected to grow rapidly, major uncertainties exist about how the sector will develop. Will it continue to make major scientific and technological progress, or will it stumble over medical challenges and controversy as did gene therapy or genetically modified foods? Will payers—such as the government, insurance companies, and HMOs—be willing to reimburse for increasingly sophisticated and more expensive devices instead of older, less costly devices or treatments? Will manufacturers be required to conduct extensive cost/benefit studies tied to the FDA's more rigorous and longer post-market surveillance requirements? How will market consolidation affect the competitive landscape and the sources of future innovation? Finally, what will be the role of foreign competitors entering the U.S. market as local companies look to expand overseas in search of future growth? Chapter 8, "Scenarios up to 2025," discusses some of the possible scenarios.

Medical diagnostics industry

The medical diagnostics industry covers laboratory testing equipment and supplies, as well as over-the-counter consumer diagnostics, such as pregnancy tests and sophisticated diabetes/blood glucose monitors.[15] Traditional descriptions of the medical diagnostics segments focus on the product type and where the test is conducted—in the hospital, in the doctor's office, or at home. But newer technologies and secure, web-based communication often blur those distinctions. Let's examine two dominant and rapidly changing segments: laboratory and point-of-care (POC) test products and services.

Major components of laboratory testing

Within the laboratory testing arena, the majority of tests fall within these three groups:

- **Clinical chemistry**—This testing covers a broad range of analyses of blood and other bodily fluids to measure critical components such as blood glucose (which can indicate diabetes), electrolytes (which can indicate kidney and other diseases), and enzymes (which can flag problems with specific organs).

- **Immunoassays**—These are chemical tests used to detect or quantify a specific substance (such as drugs, hormones, proteins, tumor markers, and markers of cardiac injury) using antibodies specific to the substance.

- **Hematology**—Blood tests show whether the levels of different substances in blood fall within a normal range. Typical tests measure levels of cholesterol, red cell/white cell counts, and blood glucose.

Laboratory testing

The U.S. market for equipment and supplies in the hospital and external laboratory testing centers is more than $45 billion today, but growing only at 1–3% per year.[16] Approximately two-thirds is for laboratory testing, with one-third for medical imaging (such as MRI and CT). The major components of laboratory testing are immunoassays (30%, all types), clinical chemistry (23%), and hematology (10%). Given the cost, complexity, and infrastructure support required for traditional laboratory testing, it's not surprising that the United States, Japan, and Western Europe account for more than 80% of the lab testing market today. However, the continuing growth of middle classes in China, India, and Latin America implies new market growth for basic laboratory services in the 15%–20% range per year, which is two to three times the rate of established markets. Not surprisingly, leading suppliers such as Abbott, GE Medical, and Siemens are aggressively expanding their overseas operations.

As discussed in Chapter 3, "Snapshot of the biosciences," advances in genetic testing are likely to radically energize, if not transform, this market. First, the traditional molecular tests for infectious diseases and blood transfusion could be replaced by DNA-based tests that are more rapid (hours versus days). Also, they offer a more extensive and deeper array of outputs, such as metrics for antibiotic resistance. Although DNA tests to assess predispositions for diseases such as cancer and Alzheimer's are still in the early stages of development, this area is causing public concern. The potential for insurance and employers to misuse such data to deny coverage or employment raises major social and ethical issues.

Point-of-care (POC) patient monitoring and diagnosis

Point-of-care (POC) products are simple to operate. They cover the range from watches that track heart rates to minimally invasive diabetes monitors. They require little maintenance and calibration, and deliver rapid, reliable results. The market for such products is estimated to be more than $11 billion, with the majority from over-the-counter U.S. sales.[17] Consumer-driven POC products range from blood pressure kits to pregnancy tests. The largest and fastest-growing component of the over-the-counter POC is for blood glucose monitoring. It is estimated to be more than $3 billion, reflecting the growing diabetes crisis, which now ranks as the fourth-leading cause of death in developed countries.

For clinical settings, POC products include various bedside diagnostic tools that are easy to transport. Examples include cardiac markers, blood gas/electrolyte analyzers, coagulation tests, and glucose monitors. Their aim is simple: to reduce the time between diagnostic test and therapeutic intervention. The major manufacturers are Abbott Labs, Roche Diagnostics, Mitsubishi, Beckman Coulter, Siemens, and several smaller technology companies, such as 20/20 Gene Systems. Their products are reducing the need for central lab tests by conducting them at the bedside. Also, they are moving out of the hospital into biodefense, field epidemiology, and other nonclinical settings. Home-based or telemedicine opportunities are a further evolution of these bedside products. For chronic diseases, such as diabetes or renal failure, "at-home robots" are being introduced that give daily medication reminders, monitor health indicators (such as weight and blood glucose), and transmit the data electronically (via phone or computer) to a centralized care group. Home-dialysis companies such as NxStage enable nephrologists to monitor their kidney-disease patients remotely and make therapy adjustments as needed.

Many challenges remain for the future. Reimbursement, patient privacy, cost issues, and standardization have limited POC applications. Today more than 50 different companies are competing in the POC market, offering more than 40 different types of tests. Information exchange by POC diagnostics, especially between existing or "legacy" hospital systems, may remain a barrier to future growth. It also might create new competition from companies such as Google, Microsoft, and Revolution Health (the online medical records company founded by Steve Case). Information exchange between devices and electronic medical records (EMR) or personal health records requires not just connectivity, but also interoperability: the ability to transfer data automatically and seamlessly. The difficulty is getting diverse information systems to communicate, such as Mac and PC platforms, or cellphones and laptops. Few hospitals today have fully functioning EMR systems or even full connectivity among their own medical equipment and central data systems. Most hospitals retain inefficient parallel systems: electronic and paper based. If Microsoft or Google, for example, can unite these disparate information systems, they could become a formidable new competitor in this growing segment.

Prevention and disease management

This final category is not normally included with the earlier four, but is increasingly important. It covers specific disease-management programs, such as for diabetes, as well as general wellness initiatives, including employer-supported fitness programs.[18] According to a 1988 Centers for Disease Control and Prevention (CDC) study, "prevention" is defined as health promotion, health protection, and preventive health services. It amounts to more than $30 billion within the United States.[19] The rationale for preventive and disease-management initiatives is simple: The United States spends the most per person on healthcare (more than $6,000 per year), while ranking low among Western nations in terms of longevity, infant mortality, and incidence of obesity and diabetes. Even worse, average annual healthcare cost increases in the United States are among the highest in the world. In an effort to lower costs and improve outcomes, prevention and disease management are receiving more attention in the United States. However, the link between behaviors and health is very complex. Although the data are overwhelming that individual behaviors can greatly impact specific risks (such as smoking and lung cancer), broader indicators such as longevity are not so easily tied to specific behaviors. For example, in Japan, nearly one-third of the population are daily smokers—one of the highest rates in the world—yet its population lives exceptionally long. And the British, who are comparable to Americans in the percentage that is overweight, enjoy a longer life span than most Americans.

Nonetheless, the rationale is strong for prevention programs and disease management of high-incidence, high-cost diseases[20] such as asthma, diabetes, and congestive heart failure. Major healthcare insurance companies such as CIGNA, WellPoint, Blue Cross Blue Shield, and Kaiser Permanente have documented success in improving outcomes and reducing costs through their aggressive wellness and disease-management efforts. Further empirical studies are needed, however, to sort out more clearly under which conditions prevention works best and why.

At Kaiser Permanente, one of the oldest and largest managed-care organizations in the United States, primary-care clinics have been reoriented to the needs of patients with chronic conditions. Specially trained, multidisciplinary care teams that include physicians, nurses, health educators, physical therapists, and others link all the components of healthcare delivery. Patients with chronic conditions such as asthma or diabetes are enrolled in support programs through aggressive outreach efforts emphasizing prevention, education, and self-management. The impacts are clear: Kaiser's integrated care system ranks very high in terms of access, cost-effectiveness, and treatment quality.[21]

Advocates for prevention and disease management also argue that such programs have a broader societal impact. For example, land-use policies supporting bike and pedestrian paths or parks built around ball fields, basketball courts, and other areas for physical activity yield indirect medical benefits for the local community A recent study of bike path cost/benefits in Lincoln, Nebraska, concluded that "for every $1 investment in trails, there was a $2.94 medical benefit in savings due to the higher physical activity levels of trail users."[22]

Conclusions

Our purpose in discussing the five key segments—pharmaceuticals, biotechnology, medical devices, diagnostics, and prevention/disease management—is to explore the weak points or "stress fractures" that could limit their role in supporting biomedical innovation in the future. In summary, we see these pressures for change in the five interrelated segments:

- **Pharmaceutical companies**—Facing major challenges in the form of weak R&D pipelines, increasing regulatory pressures, negative public perceptions, competition from generic drugs, restrictions on reimbursement, overcapacity in the sales force, and the need to better balance developed and developing markets.

- **Biotechnology**—Struggling to define sustainable business models that are distinct from the traditional pharmaceutical model which requires enormous capital and often vertical integration. Highly focused strategies developed around distinctive intellectual property, supplemented with multiple supporting alliances, present one common model for biotech.

- **Medical devices**—Maintaining physician support and guidance; expanding innovation through convergence of electromechanical, informatics, and bioscience disciplines; managing across more regulated as well as more global markets; and developing device registries in case of recall or malfunction

- **Medical diagnostics**—Managing geographic expansion of established technologies and investing in new, radically different genetic-based diagnostic testing; building public support for genetic testing; and integrating ubiquitous point-of-care test capabilities across acute and chronic care settings

- **Prevention/disease management**—Establishing standards, meeting those standards cost effectively, and integrating acute and longer-term lifestyle interventions; determining how to promote and educate people about the importance of a system's approach to health and well-being; showing efficacy through data.

How well these challenges are managed and how the underlying forces play out during the next 15 years will influence the future of the biosciences, as explored in the next chapters.

Endnotes

[1]The chapter was written jointly with Jim Austin, director of Life Sciences, Decision Strategies International, Inc., and a former senior executive at Baxter Healthcare.

[2]Lawton Robert Burns, ed., *The Business of Healthcare Innovation* (New York, Cambridge University Press, 2005).

[3]Bethany Hughes, "2007 FDA Drug Approvals: A Year of Flux," *Nature Reviews Drug Discovery* 7 (February 2008): 107–109.

[4]The "drug targets are typically enzymes, receptors on the surface of native cells or foreign antigens, or biochemical pathways or ion channels." Lawton Robert Burns, ed., *The Business of Healthcare Innovation* (New York, Cambridge University Press, 2005).

[5]Jean-Pierre Garnier, "Rebuilding the R&D Engine in Big Pharma," *Harvard Business Review* (May 2008): 68–77.

[6]A common cause of bronchiolitis and pneumonia among infants.

[7]Gary Pisano, *Science Business* (Boston: Harvard Business School Press, 2006).

[8]Lawton Robert Burns, ed., *The Business of Healthcare Innovation*.

[9]Gary Pisano, *Science Business* (Boston: Harvard Business School Press, 2006).

[10]Datamonitor, *Global Healthcare Equipment and Supplies* (London: Datamonitor Industry Reports, 2006).

[11]The device and necessary surgery to implant often cost more than $50,000.

[12]Barnaby J. Feder, "A Lifesaver, But the Risks Give Pause," *New York Times* (13 September 2008): B1.

[13]For more discussion of new materials and informatics, see the Millennium Research Group, *Year in Preview, 2007: Comprehensive Coverage of the Medical Device Industry* 2:11 (Toronto, December 2006): 31.

[14]Mike Rice, "Implantable Neurostimulation Device Market Poised for Explosive Growth," *Future Fab Intl.* 20 (7 January 2006).

[15]See the discussion of market size in Steve Halasey's article "The Global Market for Diagnostics," *IVD Technology* 5 (September 1999): 36.

[16]"Diagnostic & Medical Laboratories: U.S. Industry Report," IBIS World Industry Report (27 October 2008).

[17]Espicom, "The Global Market for Point-of-Care Diagnostics," (December 2007).

[18]Driven primarily by for-profit entities, such as Alere Healthcare, with a focus on technology-based disease-management systems for different groups.

[19]R. E. Brown et al., *National Expenditures for Health Promotion and Disease Prevention Activities in the United States* (Washington, D.C.: Battelle; Medical Technology Assessment and Policy Research Center, 1991) publication no. BHARC-013/91-019.

[20]According to K. E. Thorpe, C. S. Florence, and P. Joski, in "Which Medical Conditions Account for the Rise in Healthcare Spending?" (*Health Affairs*, 2004), 15 diseases account for 56% of our healthcare system costs.

[21]World Health Organization, "Integrating Prevention into Healthcare" *Fact Sheet #172* (14 April 2008).

[22]Prevention Institute and the California Institute, "Reducing Healthcare Costs through Prevention," Healthcare Reform Policy Brief, Working Document (2007): 11.

6

Healthcare under stress

with V. Michael Mavaddat[1]

The economic strength and vigor of the healthcare ecosystem will greatly affect the pace of innovation in medicine. In this chapter, we explore the promises and challenges of bioscience technologies in the broader context of the healthcare ecosystem. Our discussion is very selective, given the broad scope of healthcare globally. The aim is to examine major stress points in healthcare today and then identify criteria that will determine which biomedical innovations might find their way to the market, using specific case examples.

Members of the healthcare ecosystem focus on *financing* and *delivering* healthcare services, *innovating* new technologies and tools, and *regulating* different aspects of the value chain. Figure 6.1 depicts the major value-creation activities from beginning to end in healthcare delivery and offers a broad strategic overview of the complex relationships among diverse players. In the United States, different organizations compete and collaborate to meet the nation's demand for healthcare products and services. Most are operating with limited resources amid growing public concerns about the rising cost and uneven quality of healthcare.[2]

Payers provide the funding to run the healthcare value chain. They can be private entities (employers, individuals, or coalitions) or federal and state governments, such as with Medicare and Medicaid in the United States. The fiscal intermediaries, such as HMOs and insurers, perform the role of pooling risks and managing benefits for a fee. The providers, which include hospitals, physicians, and other health professionals, deliver healthcare services to patients and receive payments from financial intermediaries, the government, or patients. In the United States, distinctly different organizations deliver these three stages of the value chain. In other countries, such as the United Kingdom, the government performs the role of the payer, the financial intermediary, and the provider. In both cases, financiers who provide the investments necessary to usher in new technologies support the entire system.

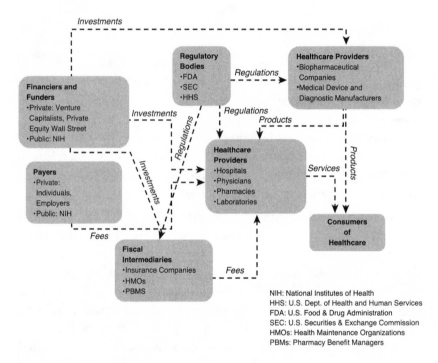

NIH: National Institutes of Health
HHS: U.S. Dept. of Health and Human Services
FDA: U.S. Food & Drug Administration
SEC: U.S. Securities & Exchange Commission
HMOs: Health Maintenance Organizations
PBMs: Pharmacy Benefit Managers

Figure 6.1 The healthcare ecosystem

Many variants of the U.K.- and U.S.-type systems exist across the globe. The approach used reflects the country's societal norms and values. In Germany, patients never need to see a medical bill. In Taiwan, a hybrid system seeks to adopt the best features of various approaches around the world. Regardless of the type of healthcare system a given nation adopts, making all technologies and innovations available to all consumers everywhere is economically unsustainable and can bankrupt governments. Healthcare economists speak of an iron triangle in which countries must make tough choices about *access* (who gets treated and for what), the *quality* of healthcare delivered, and the overall *cost* to the system (private and public).

Stress in developed nations

Most developed countries have seen their healthcare bill, as a percentage of the gross national product (GNP), rise considerably. However, interesting differences exist among countries (see Figure 6.2a), reflecting how they have balanced the iron triangle. In the case of the United States, the system has resulted in the greatest healthcare spending per person compared to other countries. Yet with more than 40 million uninsured Americans, this level of spending has not resulted in better access or better quality relative to other nations. Since the 1960s, U.S. employers have seen a major rise in their contribution to private health insurance, in addition to government-mandated taxes to pay for Medicare (see Figure 6.2b).

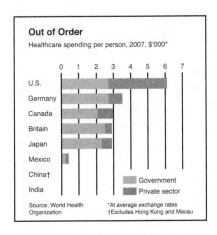
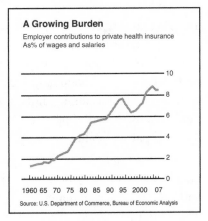

Figure 6.2 Exploding healthcare expenses

The broad use of expensive medical technologies drives much of the cost increase in the United States. Between now and the year 2025, there will be major increases in the elderly population in all age brackets. Some might be able to care for themselves and lead economically productive lives. But many will lack sufficient resources to sustain themselves for long lives and to pay for healthcare. Many of the elderly will require health services for chronic conditions and—in the absence of a cure—the goal is to help individuals maintain independence and a good level of functioning.

According to the Centers for Disease Control and Prevention (CDC), by 2030, the number of Americans with a confirmed diagnosis of dementia is expected to more than double and reach 5.2 million.[3] In addition, depression has been identified as one of the major chronic illnesses of the twenty-first century, according to the World Health Organization (WHO). These chronic conditions will burden the healthcare system even if their treatment doesn't require expensive new therapies. The concern is both the absolute size of the older cohorts and their proportion relative to the younger adults upon whom the financial burden of care will ultimately fall. The decreasing tax base will impact the availability of public resources to pay for healthcare. Given all these stresses, people are calling for an overhaul to the U.S. healthcare system, with universal access at the top of the agenda.

Healthcare funding

Tax revenue typically funds universal healthcare systems. Some nations, such as Germany, France, and Japan, use a multipayer system in which private and public contributions fund healthcare. Others, such as Canada, Sweden, and Denmark, have opted for a single-payer system, in which a single entity, typically a government-run organization, acts as the administrator (or "payer"), to collect all healthcare fees and pay all healthcare costs.

In the United States, major challenges remain for universal healthcare, including opposition by drug and device manufacturers and concerns about managing a huge universal-care program well while limiting access to legal residents. Similar to most developed countries, the United States struggles with several major challenges in its healthcare systems:

1. Expensive advances in diagnosis and treatment
2. An aging population with more chronic conditions
3. Poorly funded healthcare systems with major shortages expected
4. A deep-seated entitlement mentality about some access for all

How the nation addresses these challenges will significantly impact the availability of resources to fund bioscience technologies.

Major diseases of the developed world

Increased longevity in developed nations is accompanied by increased prevalence of chronic diseases, such as cardiovascular disease (primarily heart disease and stroke), cancer, and diabetes. According to the CDC, the United States cannot effectively address escalating healthcare costs without controlling chronic diseases, which create significant challenges for the already cash-strapped healthcare systems.

Costs of chronic diseases in the United States

- In 2005, 133 million people—almost half of all Americans—were living with at least one chronic condition.
- Chronic diseases account for 70% of all deaths in the United States.
- The medical costs of people with chronic diseases account for more than 75% of the nation's $2 trillion medical costs.
- The direct and indirect costs of diabetes are $174 billion a year.
- Each year, arthritis results in medical costs of nearly $81 billion and total costs (medical care and lost productivity) of $128 billion.
- In 2008, the cost of heart disease and stroke in the United States is projected to be $448 billion.
- The estimated total costs of obesity were nearly $117 billion in 2000.
- Cancer costs the nation $89 billion annually in direct medical costs.

Sources: Centers for Disease Control and Prevention, U.S. Department of Health and Human Services

Cardiovascular diseases

Heart disease and stroke are the two leading causes of death and disability due to illness in the United States. Heart disease causes approximately 29% of all U.S. deaths, or almost 700,000 deaths per year. Together they account for more than 20% of medical spending in the United States.[4] The major risk factors for heart disease and stroke are high blood pressure, high blood cholesterol, tobacco use, unhealthy diet, sedentary lifestyle, obesity, diabetes, advancing age, persistent inflammation, elevated homocysteine levels in the blood, and heredity or genetic predisposition.

Although angioplasty, open-heart surgery, and drug therapy are reasonably effective in managing cardiovascular diseases, they place a significant burden on healthcare budgets.

Diabetes confounds the management of heart disease and stroke. Diabetes is caused by either the failure of the pancreas to produce enough insulin (Type 1) or the body's inability to effectively use the insulin it produces (Type 2). According to the WHO, diabetes currently affects 246 million people worldwide; if left unchecked, it will adversely influence the lives of 380 million people by 2025. Adults with diabetes have heart disease death rates two to four times higher than those of adults without diabetes. The risk of death from stroke is 2.8 times higher among people with diabetes. Diabetes is also the leading cause of kidney failure and of nontraumatic amputations worldwide.

Cancer

Cancer is another disease that stresses the healthcare system. Broadly defined, cancer is characterized by uncontrolled growth and spread of abnormal cells in the human body. Cancers are typically caused by environmental factors, such as tobacco, chemicals, ionizing radiation, and infectious organisms such as viruses; or by internal factors, such as genetic mutations, hormone imbalances, and immune system dysfunctions. The environmental and internal factors often act together to initiate or promote the process of cancer development, which is called *carcinogenesis*.

In both the developed and developing world, lung cancer is the most common male cancer and breast cancer is the most common female cancer. However, some major differences in cancer rates exist between the developed and developing worlds: Cancers of the stomach and uterine cervix are higher in the developing world, but prostate and colorectal cancers are higher in the developed world.[5]

Despite the prevalence of cancer as one of the largest contributors to mortality (see Figure 6.3), medicine often remains powerless in curing this disease. Much progress has been made in elucidating the biological basis of carcinogenesis as well as by introducing preventative measures such as smoking-cessation programs. But most cancer treatments still entail invasive procedures: surgery, chemotherapy, radiation, hormone therapy, and immunotherapy. Over the last fifty years, mortality due to heart disease and stroke declined sharply, but the war on cancer remains a great challenge, especially since its incidence increases with age.

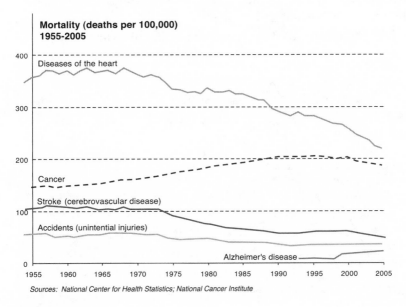

Sources: National Center for Health Statistics; National Cancer Institute

Figure 6.3 Disease mortality over time

Stress in developing nations

The biosciences might also have a dramatic impact on healthcare in developing nations, but the application of these technologies encounters major challenges. First, how can we expand these new technologies to poorer nations when the richest countries in the world, where they are first developed and deployed, have difficulties themselves controlling healthcare costs? Second, several of the new bioscience technologies (such as stem cells and genetic engineering) touch deep moral and even existential issues, such as what it means to be human. Societies around the world need to address some of these thorny ethical and religious issues before the full benefits of the biosciences can be universally applied.

Nearly two-thirds of the global population live in the developing world and try to get by with annual incomes of less than $1,500. These countries often struggle to meet basic needs: potable water, dependable food supplies, and rudimentary healthcare (such as childhood vaccination programs). The WHO reports that the mortality rates for children less than 5 years of age is 159 per 1,000 births in the least developed countries surveyed, compared to just 6 per 1,000 births in high-income countries. Figure 6.4 highlights striking scale differences in income and population size between the developed and developing worlds. In particular, note the following:

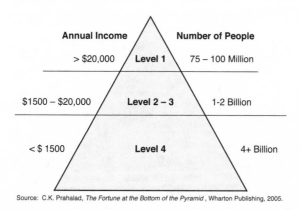

Source: C.K. Prahalad, *The Fortune at the Bottom of the Pyramid* , Wharton Publishing, 2005.

Figure 6.4   The economic pyramid

- Countries with the biggest potential consumer markets, such as China, India, Nigeria, Indonesia, South Africa, Brazil, and Mexico, still reside at the bottom of the pyramid in terms of income per capita.

- Disease incidence and progression are vastly different between the developed and developing worlds. For example, although southern Africa accounts for less than 2% of the global population, nearly 30% of the world's AIDS patients reside there.

The traditional pharmaceutical model of the developed world doesn't work well for countries at the bottom of the socioeconomic pyramid. Even diseases that affect all countries, such as cancer and diabetes, receive much greater levels of spending in developed markets. Although 16% of all diabetes and heart disease cases occur in developing countries, they hardly count in global industry revenues. For example, developing nations generate less than 2% of global sales of cardiovascular drugs.

Drug companies are often criticized for focusing on Western markets where people can pay for medicines. However, as noted earlier, the economics of developing new products for the Third World are daunting. Western pharmaceutical companies are increasingly threatened, as Abbott recently found in Brazil, where government policies forced them to either lower their prices or face government-endorsed generic competition.

Before developing nations can adopt new biosciences technologies, they need to address the following challenges:

- Government willingness or ability to finance broad, public-health initiatives
- Balancing medication costs, given the low levels of income per capita
- The lack of even basic healthcare-delivery infrastructures
- The need for simpler, more cost-effective public-health strategies, such as supplying rural communities with potable water

Even though the promise of the biosciences is high, so are the challenges. To illustrate this, we next discuss two major healthcare challenges facing the developing world: AIDS and malaria. Later we discuss how advances in the biosciences might impact them.

Acquired Immune Deficiency Syndrome (AIDS)

AIDS is caused by the human immunodeficiency virus (HIV), which was identified as the etiologic agent in 1983. It has been found in more than 200 countries and territories worldwide, and is spreading

China and India could become "laboratories of the world"

"Affordability will be the underlying factor that will drive the future of biotechnology. From now until 2020, India and China will become the key to affordable drug development. If China is the 'factory' of the world and India is the 'office' of the world, then India and China can both be the 'laboratories of the world.' India and China will outsource 'invention' from the West, and the West will outsource 'innovation' from India and China."

Kiran Mazumdar-Shaw, Chairman & Managing Director, Biocon Ltd., India

rapidly, particularly in developing countries (see Figure 6.5). Sub-Saharan Africa has the highest prevalence of AIDS among adults (16 times greater than in Western and Central Europe).

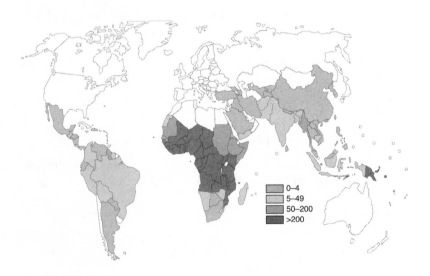

Source: Global Malaria Report 2008, WHO; in millions

Figure 6.5   Global spread of AIDS

Antiretroviral drugs, which need to be taken daily for the rest of the patient's life, provide the only effective means of keeping the

AIDS virus from destroying the patient's immune system. Although they're not a cure, these drugs can retard the progression of the disease for many years. Typically, patients need to take more than one antiretroviral drug at a time (termed *combination therapy*) to combat the rapid rise of resistance of the virus to any one drug. Another problem, especially in South Africa, is that TB and AIDS interact, requiring joint treatment.

Challenges of AIDS in developing markets

- **Cost of drugs**—The price of drug therapy is prohibitive. Consider India, where antiretroviral therapy for first-time patients costs about $25 a month at a city pharmacy—a hefty amount for many working-class Indians. India's per-capita income is approximately $250 per month.

- **Societal issues**—Cultural beliefs and perceptions, more pressing acute social issues, and budgetary problems are some of the hurdles local and national governments face in trying to combat AIDS. For example, people in some African countries believe that taking a shower after sex would reduce the risk of AIDS, or that having sex with virgins would cure the disease.

- **Infrastructure**—Most developing countries lack adequate health system infrastructures, not only to identify those infected, but also to help manage and deliver ongoing care. Although a critical demand exists for AIDS treatments, pharmaceutical companies face not only economic, but also structural barriers in delivering products to the patients.

Malaria

These problems are not unique to AIDS; they're endemic to most of the major diseases of the developing world. Consider malaria, a life-threatening parasitic disease transmitted by mosquitoes. Today approximately 40% of the world's population, primarily those living in the poorest countries, is at risk of contracting malaria (see Figure 6.6). Roughly 90% of malaria-related deaths occur in Africa; the disease is the 12th-leading cause of death in the developing world.[6]

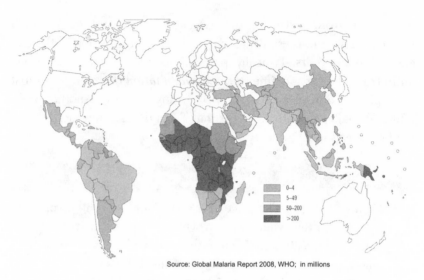

Source: Global Malaria Report 2008, WHO; in millions

Figure 6.6   Global footprint of malaria

Treatment focuses on several fronts. First, low-cost strategies appear to offer the most immediate impact. For example, mosquito nets treated with insecticide greatly reduce malaria transmission. Second, prompt access to treatment with effective up-to-date medicines (such as artemisinin-based combination therapies) reduce both mortality and lifelong debilitating effects of the disease (such as learning impairment from brain damage). Because of growing resistance among malaria parasites, these newer therapies are more effective than traditional drugs. But such combination therapies cost up to $2.40 per treatment, or roughly 10 times more than older drugs such as chloroquine, placing them out of reach of many who need the drugs.

As the examples of AIDS and malaria illustrate, pharmaceutical companies face serious challenges in applying strategies from the developed nations to the developing markets. But as middle classes grow in size and developing nations start to respect intellectual property regimes and other international standards, they could become attractive markets. With the development of domestic pharmaceutical companies in India and China, plus the development of a major bioscience initiative in Singapore, the developing world also promises to become an increasingly important center for new research, manufacturing, and commercialization. Emerging markets raise challenges of social responsibility and competition, even as they create potential market opportunities for companies that can meet local needs. Addressing the healthcare challenges of developing countries could lead to changes in therapeutic approaches, including the use of new bioscience technologies, with more emphasis on cost control.

Which technologies will succeed?

The lessons from the venture capital field are clear. Whenever a new business idea requires significant change in consumer behavior, production systems, or regulations, the odds of success decline greatly. Notable failures include Webvan.com, which offered online grocery shopping with home delivery, and various electronic medical records (EMR) systems for physician offices. Although these innovations seemed appealing at first, they required too many changes on too many fronts to succeed. In the case of Webvan, it wasn't clear where to deliver the groceries when no one was home, and many consumers really wanted to select the fruits themselves after feeling them with their hands. Likewise, many small physician offices could not afford a change from physical to electronic records because of a lack of investment, capital, scale, or trained staff. This is one reason the Obama administration wants to serve as a catalyst and orchestrator of EMR since the free market has been slow to embrace it.

The implications for healthcare are that technologies that require significant changes anywhere along the value chain will face significant hurdles and, therefore, few investment dollars. In addition, deeply entrenched positions and special interests in the healthcare ecosystem might further prevent the adoption of new technologies in the United States and elsewhere. To predict which biomedical innovations will most likely succeed, we need to assess in detail the degree of fit between the technology and the prevailing healthcare system along certain dimensions:

- **Economic and budgetary fit**—Does a budget exist to pay for the new technology or is there another way to recoup the upfront investment? Is it cost-effective?
- **Ethical or moral fit**—Does the technology align well with ethical norms prevalent in the society? Does it violate the

moral code of any significant stakeholder group (as further discussed in the next chapter)?

- **Infrastructure and distribution fit**—Does the necessary infrastructure for using the new technology and for distributing it to the point of care already exist?

- **Regulatory fit**—Is the regulatory system capable of handling the new technology through the existing channels without causing overload?

- **Business model fit**—How well does the new technology conform to the current business models of providers and other key players so that they will embrace it?

- **Medical practice fit**—Does the new technology dovetail well with practice standards and norms of healthcare professionals, from technicians and nurses to doctors?

- **Investment fit**—Does the investors' time horizon match the length of time required to develop and commercialize the new technology?

- **Requirement for scale**—Can the new technology be developed on a small scale initially for proof of principle, and then scale up to full commercialization? Is the required scale within the capacities and interests of capital markets to provide sufficient funding?

Illustrative cases

Diseases in both the developing and the developed world seem to offer ample opportunities for new biomedical technologies. And during the next 20 years, novel treatments for AIDS, malaria, cardiovascular diseases, diabetes, and cancer will emerge. However, their ultimate success—whether the local healthcare system will adopt them—will depend on how well they fit with the criteria previously defined. It will be hard to judge whether a new technology or a novel approach to treating cancer will succeed unless we can assess the specifics. The devil is very much in the details. To illustrate the framework, let's examine several promising solutions for some of the diseases discussed earlier.

AIDS and malaria

Prevention is currently the most cost-effective approach for dealing with the burden of AIDS and malaria in the developing world. It can reduce new HIV infections among men and women at high risk. Given enough time and financial resources, biomedicine might one day produce an effective AIDS or malaria vaccine. However, the widespread adoption of new vaccines could be slowed by a lack of infrastructure as well as political or moral objections. Logistical issues of getting vaccines to at-risk populations in remote parts of the developing world could also be daunting, especially if the vaccines require refrigeration or other special handling. In Africa and elsewhere, patience will be required of financiers and social entrepreneurs who want to invest in new ventures. They will entail longer payback periods, and often much lower returns, than traditional venture capitalists expect. There will be a great need for so-called *patient capital.*[7]

Cardiovascular diseases

The following list illustrates how a number of novel bioscience-based approaches might improve the burden of cardiovascular diseases (CVDs):[8]

- **Disease-modifying gene therapy**—About one third of all balloon angioplasties (a procedure to open blocked coronary blood vessels) fail because of rapid proliferation of muscle cells in the blood vessels following the treatment. This causes the vessels to close up (restenosis). To prevent this, researchers are experimenting with various genes that will stop the proliferation of muscle cells and that can be ferried to the inside of the vessels using viruses.
- **Transgenic transplant surgery**—Xeno-transplantation of pig hearts into humans could become a viable means of treating congestive heart failure and other life-threatening CVDs. Advances in biosciences could overcome rejection of foreign tissue by selectively turning off the body's immune response.
- **Angiogenesis therapy**—Novel therapies are being developed to restore blood flow to the heart, as alternatives to coronary bypass surgery, angioplasty, or clot-dissolving drugs.

Genes for various angiogenic proteins can be delivered directly into the heart utilizing gene therapy techniques. These genes trigger new blood vessel growth to the heart.[9]

- **Vaccine therapy**—If addiction to smoking (nicotine) can be prevented, a major environmental cause of heart disease could be eliminated. Several nicotine vaccines are currently in clinical development. Antibodies against nicotine can prevent this molecule from passing into the brain, thus reducing addiction.[10]

The biosciences can also deliver novel treatments for managing diabetes, which would reduce the incidence of CVDs among diabetic patients. Transplantation of islet cells (insulin-producing cells of the pancreas) that are derived from stem cells might provide a means of curing diabetes in the future.

All four therapies fit well with the prevailing business models of healthcare providers. However, by virtue of their novelty, these therapies will set new standards in the practice of cardiology and require key opinion leaders to promote them. This process of standard setting and diffusion (which also includes training in the new techniques) could mean that only specialized centers of excellence in cardiology might adopt some of these therapies. Apart from standards and expertise, cost can be a major issue. We see this clearly in the area of cancer, which we turn to next, where proton beam treatment is only offered in a few locations in the United States. The reason is simple: Each proton beam machine costs around $150 million (see the following sidebar).

Cancer

The biosciences also offer much hope of alleviating the pain and suffering associated with cancer. Many promising approaches could one day result in novel cancer treatments. Compared to traditional chemotherapy, these medications are more highly targeted to the cancer, often with fewer and more easily managed side effects. Let's review a partial list of some new approaches that are in various stages of research and development.

Proton beam cyclotrons: the most complex and expensive ever

The proton beam juggernaut has the dubious honor of being the most expensive and complex medical device ever. It is a form of external beam radiotherapy in which protons, instead of photons or electrons, are used to irradiate the tumor. The advantage of using protons to deliver ionizing radiation is that the shape of the beam can be controlled with extreme accuracy to conform to the shape of the tumor being treated. Protons do not scatter much in the body given their larger mass, so the beam stays focused on the tumor which limits radiation exposure to the surrounding healthy tissue. This complex medical device was co-developed by physicists who smash elementary particles into each other near the speed of light in huge accelerators. The proton cyclotron can fire multiple beams simultaneously into the tumor with great accuracy. The protons loop around the internal accelerator at least 10,000 times before unleashing their lethal force. The treatment is brief—the patient often walks out in minutes.

The machine itself is enormous, using about 1,600 superconducting magnets which produce kinetic energy equivalent to a train traveling at 100 miles an hour. The University of Pennsylvania hospital is developing the largest proton therapy institute in the world at the Perelman Center for Advanced Medicine. This cyclotron will be housed in a building the size of a soccer field. It weighs around 220 tons and will deliver five separate sub-beams to accommodate patients in different rooms of the hospital. Each treatment room has 18-feet-thick walls with heavy equipment weighing 90 tons that can fully rotate around the patient. The heart of the accelerator is underground, as the CERN collider in Switzerland is.

At the end of 2008, there were various proton therapy centers in Canada, China, England, France, Germany, Italy, Japan (5 centers), Korea, Russia, South Africa, Sweden, Switzerland, and the United States (6 centers)—a global total of 26 installations. Another 26 centers were under construction (at the time of this writing) in 2009. Over 60,000 patients have already been treated for solid tumors in the prostate, lung, neck, breast, and other organs. The overall cost benefit ratio of proton therapy still has to be compared to surgery, traditional radiation, and other methods commonly used for larger patient populations.

Preventative and therapeutic vaccines

The human papilloma virus (HPV), which is spread through sexual contact, can cause cervical cancer. Merck and GlaxoSmithKline make vaccines (Gardasil and Cervarix, respectively) to protect against this cancer. These HPV vaccines are *preventative*, as is the cancer vaccine against Hepatitis B (HBV) virus. Chronic infection with this virus can lead to liver cancer. Today most children in the United States are immunized against Hepatitis B shortly after birth.

Newer vaccines are being developed as therapies or treatments for cancer. These *therapeutic* vaccines prime the patient's immune system to recognize and kill cancer cells. This is no easy feat: cancer cells are mutations of the patient's own cells, so the immune system often does not recognize them. Several types of cancer vaccines are being researched, such as:

- **Antigen vaccines**—These vaccines use tumor-specific antigens that are normally displayed on the surface of tumor cells. When injected, they prime the immune system to produce specific antibodies (single proteins that neutralize antigens) and killer T-cells that recognize and destroy cancer cells that display the antigen. Weakened or killed cancer cells that carry a specific cancer antigen can also be used to make cancer vaccines. These can be taken from the patient's own cancer.

- **Dendritic cell vaccines**—These vaccines consist of human dendritic cells, which play an important role in activating the immune system against foreign intruders, such as viruses. To prepare cancer vaccines using dendritic cells, these cells are isolated from the patient's blood, exposed to the specific tumor antigen in vitro, and then infused back into the patient, where they can activate the immune system to attack tumor cells.[11] Sipuleucel-T (Provenge, Dendreon Corp) for prostate cancer is a dendritic cell-based vaccine which has shown promising results in phase 3 clinical trials.[12]

Monoclonal antibodies

Highly selective monoclonal antibodies that specifically block or neutralize cancer targets are becoming important cancer therapeutics. The following are monoclonal antibody cancer therapies:

- **Inhibitors of angiogenesis**—In 1971, Judah Folkman proposed that tumors rely on new blood vessels (angiogenesis) to survive and grow. He led the discovery of vascular endothelial growth factor (VEGF), which spurs new blood vessel growth, as well as the discovery of angiogenesis inhibitors. The drug Avastin, which blocks the VEGF pathway, was approved for colorectal cancer in 2004. Unfortunately, as with so many cellular pathways, angiogenesis relies on redundancy, using at least 12 other pathways that Avastin has no effect on. Currently, dozens of other angiogenesis inhibitors are in development, and hopes remain high for this type of therapy.[13]

- **Herceptin**—The Her2 receptor resides on the surface of certain breast cancer cells and provides a specific attachment site for growth factors that spur cell proliferation and growth. A specific monoclonal antibody can attach to this molecule and prevent growth factors from attaching, thus preventing growth of the cancer. In 1998, the antibody Herceptin was approved for use in breast cancers susceptible to Her2-fueled growth. Other targeted therapies in clinical use are Erbitux for treatment of colorectal cancer and Rituxan for treatment of non-Hodgkins lymphoma.[14]

- **Targeting obesity**—Drugs are being developed to target the insulin-like growth factor 1 (IFG-1), a close relative of insulin. One theory is that excess weight gain can set off a cascade of molecular events that leads to increases in insulin and IFG-1. Both of these hormones can then spur cell proliferation and tumor development by binding to and activating the IGF-1 receptors on cells. Pfizer is developing a targeted antibody that blocks the function of IGF-1.[15]

Chapters 2, "A short history of biomedicine," and 3, "Snapshot of the Biosciences," provide more general information about monoclonal antibodies.

Gene silencing

Several companies are working on gene-silencing techniques using RNA interference (RNAi), a technology explained in Chapter 3. These companies are designing sequences of small RNAs that are complementary to the mRNAs of known cancer genes and can block the activity of these genes. They are also working on delivery systems, such as specially designed viruses that can transport these small RNAs into the target tissue in vitro.[16]

Controlling inflammation

Inflammation is an important part of our innate immune response. It springs into action when an injury occurs or when foreign invaders are detected. It is increasingly recognized as an important contributor to nearly all chronic diseases, such as rheumatoid arthritis, diabetes, heart disease, and cancer. However, precancerous tissue has many of the same cells and chemicals that are found in the innate immune response. The inflammation response might thus inadvertently promote cancer to the next stage, where it starts to invade surrounding tissues and becomes malignant. The immune system is a double-edged sword, sometimes helping to promote cancer and other times defending against it. Researchers are looking for strategies to halt cancers by targeting cells involved in inflammation. Such medicines as the nonsteroidal anti-inflammatory drugs (NSAIDs), aspirin, and COX-2 inhibitors could be given preventatively to high-risk individuals. Drug companies are developing even more selective inhibitors to control inflammation as it relates to cancer.[17]

Stem cell approaches

Stem cells might be the main reason cancer is so hard to completely eradicate. Stem cells have been found in many types of cancers, including breast, colon, prostate, and brain cancers. Radiation and chemotherapy target actively dividing cells, but cancer stem cells generally do not divide, providing an explanation for why treatment can destroy most of a tumor, yet the cancer comes back. Cancer stem cells are also better than regular cancer cells at repairing damage to

DNA caused by radiation, making cancer stem cells a major target for pharmaceutical and biotech companies.[18]

A common impediment to the broad adoption of these therapies is the lack of funds, especially considering the high cost of these treatments. The retail price of anti-angiogenesis drugs for a one month's supply can run as high as $10,000. For treatment to be effective, a patient must be on these drugs for at least four months, which puts these medications out of the reach for those who don't have health insurance that covers it. These therapies must also fit well with the other criteria for adoption listed earlier.

Figure 6.7 illustrates that different therapies for these diseases entail distinct profiles, based on our own subjective assessments. The degree to which a given technology fits with the ideal profile determines how readily the technology becomes a commercial success.

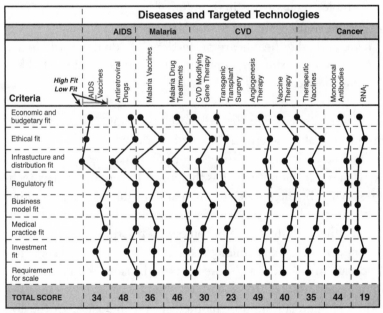

The technology is rated on a scale from 1 (low fit) to 7 (high fit) for each criterion. The total score is the sum of all eight ratings. The maximum total score is 56, representing the highest degree of fit across all eight criteria.

Figure 6.7 Technology profiles

Summary

This chapter illustrated how and why the broader healthcare ecology in a nation will greatly influence the kind of biomedical technologies that will have the most impact in the future. After discussing some of the major stresses facing healthcare systems today in both the developed and developing worlds, we presented a framework for assessing a new technology's chance of success. Because various conditions must be met for new technologies to be economically viable, many promising ones won't succeed. In general, successful technologies will need to offer significant enhancements along at least one of the following dimensions:

- **Cost savings**—Will the new technology produce the same healthcare outcome and quality at a lower cost (such as offered by various generic drugs)?
- **Quality improvement**—Will the technology offer a higher quality of life, with proven clinical outcomes, perhaps at somewhat higher overall cost (as in the case of anti-angiogenesis therapies such as Avastin or Sutent)?
- **Productivity gain**—Will the new technology enable physicians and other healthcare professionals to do more with less in terms of time, staff, and equipment (as, for example, with electronic medical records)?

As healthcare systems around the globe evolve, they will likely redefine which of these criteria to weigh more, which will impact the type of technologies that will succeed or fail. In each case, the healthcare system will influence who has the power to choose the new technologies. Will it be the physicians, the patients, the insurers and other financial intermediaries, employers, the marketplace, or the government? Healthcare systems can become either bottlenecks or accelerants for new bioscience technologies. However, healthcare ecologies don't operate in isolation. They are part of a broader economy, which itself is shaped by global trends and uncertainties that indirectly influence the commercialization of biomedical technologies. In the following chapter, we explore these broader factors, which relate mostly to the macro-environment around the globe. This sets the stage for different scenarios that new bioscience technologies might encounter in years ahead.

Endnotes

[1]This chapter was written jointly with Michael Mavaddat, a senior partner with Decision Strategies International. He has worked extensively as a healthcare and life sciences consultant, and he earlier worked as a line manager at Bristol Meyers Squibb Company. Michael is a Senior Fellow at Wharton's Mack Center for Technological Innovation.

[2]For further details, see Lawton R. Burns, *The Health Care Value Chain* (San Francisco: Jossey-Bass, 2002), and Michael E. Porter and Elizabeth Olmsted Teisberg, *Redefining Health Care* (Boston: Harvard Business School Press, 2006).

[3]D. J. Thurman, M.D., Centers for Disease Control and Prevention, unpublished data, 2006.

[4]You can find these and other cited statistics on the websites of the Centers for Disease Control and Prevention in Atlanta, Georgia, and the World Health Organization in Geneva, Switzerland.

[5]Raymond W. Ruddon, *Cancer Biology* 4th ed. (New York: Oxford University Press, 2007).

[6]*Global Malaria Report 2008*, World Health Organization. The chart shown is from Figure 3.3 of this report, on p. 33. Available at www.who.int/malaria/wmr2008/malaria2008.pdf.

[7]Thomas L. Friedman, "'Patient' Capital for an Africa That Can't Wait," *New York Times* (20 April 2007). Available at http://select.nytimes.com/2007/04/20/opinion/20friedman.html?hp.

[8]*The Atlas of Heart Disease and Stroke*. Available at www.who.int/cardiovascular_diseases/resources/atlas/en/.

[9]American Heart Association, "Gene Therapy." (May 6, 2009). Available at http://www.americanheart.org/presenter.jhtml?identifier=4566.

[10]Patrik Maurer and Martin Bachmann, *Vaccination Against Nicotine: An Emerging Therapy for Tobacco Dependence*. Available at www.ingentaconnect.com/content/apl/eid/2007/.

[11]Patrick Barry, "Combined with drugs, vaccines against tumors may finally be working," *Science News* 172 (August 11, 2007) p. 88-89.

[12]Martha Kerr, "AUA 2009: Prostate Cancer Vaccine Significantly Improves 3-Year Survival," American Urological Association 104th Annual Scientific Meeting: Late Breaking Abstract 9, presented April 28, 2009.

[13]Douglas Hanahan, and Robert A. Weinberg, "Judah Folkman (1933-2008)," *Science* 319 (Feb 22, 2008) p. 1055.

[14]Sharon Begley, "We Fought Cancer..." *Newsweek* (September 15, 2008) p. 43-66.

[15]Laura Beil, "Weighty evidence: The link between obesity, metabolic hormones, and tumors brings the promise of new targets for cancer therapies," *Science News* 173 (February 16, 2008) p. 104-106.

[16]Fumitaka Takeshita and Takahiro Ochiya, "Therapeutic potential of RNA interference against cancer." *Cancer Science* 97:8 (2006) p. 689-696.

[17]Gary Stix, "A malignant flame," *Scientific American* (July 2007) p. 60-67.

[18]Jean Marx, "Cancer's Perpetual Source?" *Science* 317 (August 24, 2007) p. 1029-1031.

7

Wildcards for the future

Thus far, we have described the biosciences as they relate to health-care, examined their relationships to other industries such as information technology, and reviewed the current business segments through which the biosciences are commercialized. We have also discussed the stresses on healthcare systems around the world. We now turn an eye to the future, looking out to about 2025. This year is more than just an arbitrary marker of a quarter century; it represents a time period that is especially relevant to us and our children. It also falls within the planning horizon of public and private organizations that will help shape the future. Yet 2025 is long enough to move our thinking about the biosciences beyond just incremental change.

Although trying to forecast the future precisely is tempting, acknowledging that much is uncertain is more honest intellectually. The problem with future projections is that they are often wrong, especially when it comes to technology forecasting. Sometimes the projections are overly optimistic, as happened with artificial intelligence and supercomputing several decades back (the so-called Fifth Generation project of MITI in Japan), or unduly myopic in not seeing the swift impact of technology, as with fax machines, personal computers, and cellphones. To avoid the trap of unbridled fantasy on one hand and the danger of incremental thinking on the other, we favor a scenario planning approach.

Trends versus uncertainties

Scenario planning is an established methodology for envisioning different possible futures, and companies and governments alike have used it to great effect.[1] The Rand Corporation was an early devotee of scenario planning in the United States, and Royal Dutch Shell used it extensively in the early 1970s to anticipate various oil shocks and other geopolitical upheavals. The basic idea behind scenario planning is quite simple. We examine the many forces that might shape the future and put them into two categories:

1. Long-term trends that we can validate and extrapolate from where we stand today
2. Key uncertainties that are less clear and subject to wide outcome swings

Whether a force, such as the aging population in developed countries or the use of stem cells in medicine, is a trend or an uncertainty hinges on empirical data, forecasting models, and expert consensus. Our own rule is simple: If experts disagree strongly about whether a particular force is a trend, we deemed it an uncertainty. In this chapter, we discuss some of the major uncertainties, or wildcards, that might significantly impact the biosciences. We discuss them mostly in isolation, to convey why each is relevant to the future of biomedicine. Thereafter, we blend these wildcards into different scenarios.

The following list summarizes various long-term trends that will influence the biosciences as well. By definition, trends are present in every scenario, as a common backdrop. What sets scenarios apart are the uncertainties, especially the wildcards. The following trends were already mentioned in earlier chapters or are well known, so we do not discuss them further.

Society and Politics

   Aging population in developed nations
   Increased use of illegal drugs and medicines
   Enhanced threat of bioterrorism
   Increased need for education and training

Science and Technology

   Continued explosion of technological innovation
   Systems biology and convergence of technologies
   Increasing concerns about global warming
   Customization of treatment, personalized medicine

Business and Economics

> Food shortages and related inequities
> Rising cost pressures in healthcare
> Globalization of bioscience research

Wildcards

Unlike trends, which move along a path that we can see and extrapolate in terms of direction, *uncertainties* entail issues that cannot be reliably predicted at this point in time. We call them *wildcards* because each can rock the biosciences:

Societal and Political

> Public acceptance of the biosciences?
> A major new disease crisis or pandemic?
> Impact of biotech rogue states?

Science and Technology

> Raging success or major meltdown?
> Role of complementary industries?
> Intellectual property regimes?
> Qualified staff for healthcare and research?

Business and Economics

> Economic growth and global power shifts?
> Venture capital and other funding sources?
> Climate change and resource scarcities?

Society and politics

In this section, we discuss some major wildcards relating to politics and society, recognizing that additional uncertainties exist.

Public support of the biosciences?

Public *acceptance* includes a favorable legislative or investment climate, generous levels of medical cost reimbursement, enthusiasm by students to work in a particular area of bioscience, or the desire of a government agency to research a disease affecting a large segment of the population. Public *rejection* can range from organized opposition by advocacy groups that try to prevent the use of a particular technology, to general apathy by a public that simply might not care whether a particular technology is developed. Active rejection has been seen in abortion technologies, human cloning, embryonic stem cells, genetic engineering, xeno-transplantation (transferring animal organs into humans), and genetically modified organisms (such as genetically engineered crops and livestock). Despite these bans, each of these technologies is being researched somewhere in the world. Whether the public accepts or rejects bioscience technologies depends on a variety of factors, from social and political to economic and technological.

Public support for bioscience research can shift over time, depending on public sentiment, the efforts of lobbyists, government influence, and other factors. At one time, the space program was the hot topic in government and research circles, followed by the computer and IT revolution. In the late 1990s, the new hot topic was human genomics and dotcoms. In recent years, nanotechnology has taken the spotlight, and now green technologies are much in vogue. The research community is also still excited about the potential of genomics to provide new biomedical therapies for serious diseases. But what happens if interest and resources shift into another area altogether, such as global warming? These kinds of shifts, which could be short term and largely unexpected, could jeopardize bioscience research, which requires consistent, long-term commitments to succeed.

The potential for rejection is real. Certain countries ban various bioscience technologies, from embryonic stem cell research to human cloning. Opposition to the use of animals in medical research also is rising. Animal rights organizations, such as the U.S. Humane Society, PETA (People for the Ethical Treatment of Animals), and the World Society for the Protection of Animals are active in political lobbying. They have organized protests and even launched attacks on research labs (although some of the more violent actions have created a backlash against these groups). The debate over how to use animals in scientific research will continue, as the bioscience community attempts to balance the need to develop life-saving drugs for humans with the

Strong opposition to Oxford biomedical sciences

In fall 2003, Oxford University began construction on a new £18 million animal-testing facility for biomedical research in the United Kingdom. Demonstrations against the construction were spearheaded by the British animal rights group, *SPEAK, the Voice for the Animals,* which had played a key role in an earlier protest against primate research at Cambridge University. In July 2004, construction came to a complete halt when the contractor pulled out amid threats and intimidation from protestors. Despite the stop, tensions continued to rise. The government came to the university's support in November by issuing a court order barring protestors from a 50-yard exclusion zone around the construction site. Construction resumed by the end of the year, under a new contractor. But conflicts flared again in January when the Animal Liberation Front, an activist group, told its supporters to target any academic, student, or company affiliated with Oxford.

Students and scientists supporting the laboratory's construction hit back by staging a march in February 2006. In this counter-protest, Oxford professor John Stein was quoted in the *Times:* "I feel passionately that animal experiments have benefited mankind enormously, and almost all of the medical advances of the last 100 years have happened through animal experiments. People just don't seem to know this." Later that year, two improvised bombs exploded in a sports pavilion. After five years of protest, the laboratory opened its doors in November 2008. Oxford claims that the facility provides better living conditions for laboratory animals than older facilities did. SPEAK is opposed to *all* animal testing.

ethical considerations of how animals are used in research (see the accompanying sidebar).[2]

A major pandemic?

No one can predict which new public health challenges will arise next. But health officials and scientists consider it a matter of *when*—not *if*— another global epidemic of the stature of AIDS or a global pandemic, such as avian or swine flu, will strike. The SARS and avian flu outbreaks in Asia during 2003–2006 and the swine flu pandemic that began in the spring of 2009 have created a great deal of anxiety about the prospects for future pandemics. Although none of these have caused large numbers of human casualties (at least not to date), they raise fears about the possibility of another pandemic with the lethality of the 1918 Spanish flu that killed 50 to 100 million people worldwide. Scientists are concerned because humans are encroaching ever more deeply into areas of the globe that were once the province of animals. In poorer regions of Africa, humans consume meat from wild animals, such as monkeys and baboons, which brings them in contact with animal pathogens for which humans have no immunity. This happened in the case of the AIDS virus, which was acquired by humans in Africa who consumed meat from chimpanzees. In parts of overpopulated Southeast Asia, humans live in close contact with chickens and pigs, which promotes the exchange of viruses and the eventual emergence of new respiratory diseases such as influenza and SARS. Climate change is further exacerbating this problem by driving humans from once bountiful agricultural lands and coastlines in search of new sources of food.

> "Pandemics are like hurricanes, tsunamis, and earthquakes. We had 10 of them in the last 300 years and we're due for another one sometime soon."
>
> *Michael Osterholm, Director of the U.S. Center for*
> *Disease Research and Policy*

Biotech rogue states?

The potential exists that rogue states could act as a safe haven for technologies such as human cloning or bioweapons that many civilized nations deem highly unethical or excessively dangerous. The world community might not be able to control nations that pursue unethical, dangerous, or uncontrolled experiments. We have seen these challenges arise with rogue states and nuclear weapons technology (notably, Iran and North Korea). Nations with weak central governments, such as Pakistan, might not be able to stop flourishing trade in nuclear bomb technologies (as has already happened with Dr. Khan's notorious black market network). After the end of the Cold War, the former Soviet Union revealed some of its extensive biowarfare programs, which included the weaponization of smallpox and other diseases. Whether these initiatives have been cancelled, continued, or diverted is unknown and presents a future uncertainty.

The 9/11 attacks on the United States underscored the reality that small, highly motivated and well-organized terrorist groups can cause great harm and suffering. Even individuals acting alone or in small groups can use dangerous technologies to disastrous ends. Examples include the terrorist group Aum Shinrikyo in Japan, which spread Sarin gas in a Tokyo subway, and the unknown individual(s) that delivered highly poisonous anthrax spores via the postal system. In September 2001, letters containing anthrax were sent from Trenton, New Jersey to several American television and newspaper offices. A second mailing containing "weaponized" anthrax was sent to the offices of two U.S. senators. These attacks infected more than 20 people and resulted in the death of Robert Stevens, a photo editor of a Florida tabloid.

Science and technology

In this section, we discuss some major wildcards relating to science and technology, recognizing that additional uncertainties exist within this domain as well.

Raging success or major meltdown?

What are the odds that we will see a biosciences "meltdown" similar to the Three Mile Island nuclear power disaster? Failure of a widely adopted medical device or tragic side effects of a new bioscience therapy could be a showstopper. This happened a few years ago in the United States when a genetically engineered adenovirus, used in a gene therapy experiment to treat a rare metabolic disorder, triggered a massive immune response that killed an 18-year-old patient (see Chapter 3, "Snapshot of the biosciences").[3] This put a damper on gene therapy research for years. The uncontrolled spread of a genetic mutation from a genetically engineered crop could have a similarly dampening effect on bioscience research.

Conversely, a killer application could be found that elevates all biomedical research—similar to the convergence of computing, databases, browsers, and telecom that spurred the incredible rise of the

Internet, laptops, and cellphones. It could be a device that measures cellular health daily, a stem cell treatment that cures diabetes, or an emergency health kit that no family wants to be without. Success in the biosciences might take years or decades to achieve, but when it occurs, the impact can be enormous. For example, when scientists discovered that most cervical cancers are caused by the human papilloma virus (HPV), they developed effective vaccines (Cervarix and Gardasil) to prevent this deadly cancer.

Many other areas of bioscience hold great promise, but we still have much to learn about the role of genes, the molecular causes of disease, human brain function, and more. Successes will come, but where, when, and in which areas of medicine remain highly uncertain.

Role of complementary industries?

Complementary industries can have a strong supporting impact on a newly emerging technology. For example, the computer revolution could not have happened without support from related industries such as telecom, consumer electronics, cable networks, information vendors, and the media. A great deal of cross-fertilization among industries prompts some surprising innovations. In the past decade, for example, Global Positioning System (GPS) and cellphone technologies enabled General Motors to develop OnStar, the onboard navigation and interactive system that a car owner can subscribe to for a monthly fee. The need to identify criminals at crime scenes has stimulated research in DNA identification.

In many industries, scientists are exploring the interface of nanotechnology and biology. Bioengineering researchers are combining mechanical engineering and medicine to provide artificial limbs. Electrical engineers are experimenting with brain-machine interfaces that might help paralyzed accident victims. In the future, human healthcare could benefit in surprising ways from new types of supercomputers, wireless technologies that enable remote medical care, and nano-miniaturization. We examined the convergence of different industries, and the associated crossover impacts, more fully in Chapter 4, "Bio-driven convergence."

Intellectual property regimes?

The ability to protect a new drug or therapy can be a major uncertainty. The typical patent life for a prescription medicine in the United States is 11–12 years, compared to an average of 18.5 years for other patented products. After a patent is granted, a product might experience a delay in getting to market, the manufacturer might be challenged in court for patent infringement, or competitors might launch a competing drug to replace the original. In 1984, Congress passed the Drug Price Competition and Patent Term Restoration Act, which enables generic copies of medicines to quickly enter the market when safety and effectiveness has been established. In 2005, one of every two drugs prescribed was a generic compound.

Uncertainty also arises over how patents will be awarded in the future for genes and related biological components. Patent abrogation in some countries is another major area of concern for bioscience patent holders. A related example is the South African government's violation of intellectual property (IP) concerning AIDS drugs (and subsequent settlement with major pharma companies), earlier this decade.

The uncertainties posed by patent protection and expirations, as well as competition from generic drugs, have a direct impact on recovery of research costs (which can exceed $1 billion for a blockbuster drug), as well as the selection of research targets and research partners. How the IP protection landscape will evolve is a major uncertainty for bioscience research and development. Will developing nations honor international patent agreements? What kind of biological agents and processes will be patentable in the future? To what extent will courts and societies enforce IP rights?

Qualified staff for healthcare and research?

The United States, China, and many other countries have predicted major shortages of healthcare professionals. As early as 2004, the National Science Board reported a "troubling decline" in the number of U.S. citizens training to become scientists and engineers, creating an "emerging and critical problem" for this country. The U.S. Department of Health and Human Services estimates that, by 2020, the nation will need 2.8 million nurses, 1 million more than the projected supply. Concern also is growing that there might not be enough educators and research professionals to keep pace with (and sustain) rapid innovations in the life sciences. The next decade will reveal whether academia, science, and industry will be able to keep pace with the need for trained professionals. Although the demand for qualified people creates a trend (the rising need for education), uncertainty remains about how this need will be met. A silver lining for the United States is that continued outsourcing of clinical and administrative tasks to offshore countries may actually free up some local supply of labor. Also, skilled labor is being imported.

Business and economics

In this section, we discuss some major wildcards relating to business and economics, recognizing again that additional uncertainties exist.

Economic growth and global power shifts?

An important macroeconomic uncertainty is the extent to which the G-8 nations will continue to exert economic leadership in the world economy during the coming 15 years. Will increased influence from the rising economies of India, China, and other nations redraw the world economic map? The liberalization of economies in Brazil, China, India, Russia, and the former Soviet republics provides new markets for bioscience technologies and new sources of labor and research support. Entrepreneurs and investor-funded ventures in those countries may increasingly pursue biomedical opportunities. But how fast will these economies develop, and when will they shift the balance of power globally?

The world economic order, driven by globalization across a broad front of technologies, alliances, and trade agreements, might encounter deep structural problems, such as rising oil prices, overpopulation, pollution, global debt, and trade deficits. The 2008–2009 U.S. housing and credit crisis spread quickly to the global banking sector and then to the main economy, causing serious recession in many countries. Global power shifts will continue to occur. From 1985 to 2005, we witnessed the breakup of the Soviet Union, the economic liberalization of India and China, the consolidation of the European Union, and a new global connectedness through the Internet and telecom infrastructures. Global terrorism also rose in the past decade.

Venture capital and other funding sources?

The U.S. biotech industry endured a massive revaluation and restructuring to survive in 2002. By the end of 2002, market capitalization for the industry had fallen 41%. After the "biotech bubble" burst, the industry rebounded and biotech investing recovered, but these fluctuations represent a major area of uncertainty and concern. In 2008 and 2009, the global collapse of financial credit institutions posed new threats to firms that depend on venture capital and bank financing.

As the world recovers from economic forces that have impacted the fluidity and availability of capital, how will the bioscience industry fare between now and 2025? Investments in many areas of biomedicine have the potential to rise and fall dramatically. New business and research models will impact how bioscience research is funded and developed in the decade(s) to come, but the availability and role of venture capital remain uncertain, especially as green technologies vie for funding as well.

Climate change and resource scarcity?

Although global warming was listed earlier as a trend, it is also a major uncertainty in terms of its speed and impact. Devastating tsunamis and hurricanes, droughts in Africa and other regions, melting of the polar ice cap, widening of the ozone hole, and other indicators suggest that climate change will remain an area of great uncertainty for the global village. These catastrophic events pose special problems for businesses and governments, and will likely cause major social, political, and even military upheaval. The indirect effects of climate change are especially hard to ascertain because they can exacerbate the spread of disease, starvation, migration, conflict, and war.

Some of the consequences of climate change will pose direct challenges for bioscience research and development. For example, global warming might cause the migration of diseases from tropical or remote regions to populated temperate regions. Large urban populations in the industrialized nations may require vaccinations against new diseases that were previously foreign to the Northern Hemisphere. More ominously, the pressure on natural resources could prompt further regional and global conflicts, including perhaps a resurgence of bioterrorism by those in dire straits. This in turn will prompt the development of novel vaccines or medicines to counter the effects of bioweapons.

Global water use has increased sixfold during the past century. The United Nations predicts that, by 2025, half the world's population will face serious fresh-water shortages. In the developing world, 80% of water is used for agriculture, a proportion that is not sustainable. The water table in many major grain-producing areas of northern China is falling at a rate of 5 feet per year, and water tables throughout India are falling about 3 to 10 feet per year. In the next few decades, many countries will need to address water shortages caused by population growth, global warming, drought, pollution, and other factors. For example, China has 22% of the world's population but only about 8% of the world's freshwater—and China's population is growing. China's South–North Water Transfer Project will divert billions of gallons of water from several major rivers in the water-rich south to the poorly irrigated north. This is the largest water-transfer project ever attempted and is an example of the scale of projects needed in many regions to address the looming global water crisis.

Each year, drought is blamed for mass starvation, death, and disease among millions of people throughout Africa. Global warming will exacerbate this situation. Bioscientists are developing genetically modified crops that are tolerant of drought and salt, which could relieve the demand on fresh water for irrigation. However, many countries do not accept genetically modified organisms (GMOs), so the impact that this bioscience innovation will have on water preservation remains uncertain.

A scenario framework

Based on the macro forces discussed thus far, we have identified two overarching themes that we think will most shape the future of biosciences: technological success and societal acceptance. Using these two broad uncertainties, we have created a framework to describe four different scenarios we could see between now and 2025.

The first overarching uncertainty: technological success

The most important uncertainty lurks in the laboratory as well as in the human body. Will bioscience research deliver—and how quickly? Will it be like the raging success of the Internet or more like the dashed hopes of fuel cells or clean energy? Bioscience technologies offer the potential for longer and healthier lives with treatments for debilitating and fatal conditions. Also possible, however, is that these promising technologies will evolve much more slowly than expected or fail to achieve critical mass by 2025. Many will recall the optimistic projections about the war on cancer decades ago, and the formidable challenges that still remain.

> **If technology succeeds**—New therapies will likely stem from deeper analyses and manipulations at the molecular level of organs, tissues, cells, genes, and proteins, as well as from the convergence of bioscience technologies and systems biology with information technology, telecommunications, and nanotechnologies. We can expect new types of sensors, implants, drug-delivery systems, and artificial organs, along with improved diagnostics to detect diseases before symptoms appear. These emerging life science technologies will enable new forms of personalized medicine.

> **If technology falters**—Stem cell research, tissue engineering, and gene therapy breakthroughs could be less dramatic than anticipated, or a stunning failure could seriously stifle genetic research. For example, if lethal viruses or bacteria that are resistant to known cures emerge from research laboratories, we could see unprecedented pandemics. Weaponized pathogens in the hands of bioterrorists could likewise wreak havoc worldwide.

The second overarching uncertainty: societal acceptance

As we saw in the intense debates over genetically modified foods, cloning, and stem cell research, many societal forces could derail the development of promising technologies. Public opinion will influence where limited funds are allocated to benefit the greater public good. Advocacy groups, religious organizations, legislators (responding to their constituencies), and other societal groups will weigh in on the types of bioscience research conducted. Each of these groups can exert a positive or negative impact on the eventual success of the biosciences.

If strong public support emerges—Emerging bioscience technologies could benefit substantially from supportive legislation, increased investment in R&D, and favorable media coverage. More voter-supported initiatives will emerge, as with the California proposition to fund stem cell research. More countries will pass laws liberalizing the use of genetics, especially for research and the subsequent commercialization of drugs or devices. Many countries will establish well-funded research centers, as has already happened in Finland, Ireland, Japan, Korea, Singapore, India, and Italy, among others.

History shows that seemingly incurable diseases can benefit from public efforts to raise research funds and influence government action. Strong public support for AIDS research helped turn a fatal disease into a chronic one that for many patients can remain stable for decades with medication. The Human Genome Project is another notable example of what can be accomplished with strong public support and government-industry alliances.

If strong public opposition emerges—Innovation could be slowed or even halted. In the 1990s, many countries banned genetically modified organisms and related technologies prompted by the "Green Movement" and anti-GMO organizations, particularly ones in Europe. In the United States, no new nuclear power plants have been commissioned for many decades now, the result of both the escape of radioactive material at Three Mile Island in 1973 and, later, in 1986, a major nuclear meltdown at Chernobyl in the Soviet Union. If major segments of the public opt out of promising technologies (as happened with genetically modified foods, cloning, embryonic stem cell research, and abortion), entire areas of research and medicine may be stalled, delaying medical solutions for possibly millions of needy people.

Multiple scenarios are possible

When faced with multiple uncertainties, we can either try to examine all combinations or explore a more limited set in detail to get a feel for the range of futures that might lie ahead. Scenario planning is a tool to do the latter. As Figure 7.1 shows, we can think of the future as an uncertainty cone that gets wider as we look farther into the future. Akin to the weather cones we see on television when a serious storm or hurricane is threatening coastal regions, the direction and width of the cone marks the area of potential impact. The farther we look into the future, the wider the cone gets. The wide ellipse at the end of the cone denotes the range of possible bioscience scenarios for the year 2025. The future most likely will end up at some point within this contour, as suggested by the dashed curve in the chart. Even though we can't know exactly where the dashed curve will end up inside the large ellipse, we can define a range of scenarios (such as points A, B, C, and D). These scenarios are no more than detailed descriptions of selected spots at the boundary of the cone, and together they bound the range of possible futures for the biosciences.

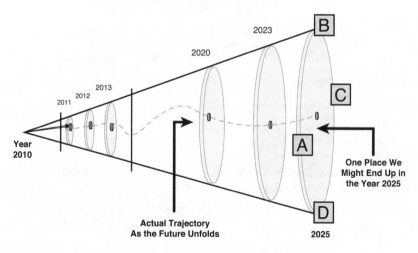

Figure 7.1 Uncertainty cone of the future

To build our scenarios, we start with the two overarching uncertainties discussed earlier: the progress of technology and public receptivity. Crossing these two uncertainties in a grid (see Figure 7.2) creates an organizing framework with four rather different scenario themes. The four cells of this grid are the starting points for developing more detailed scenarios. Specifically, the other uncertainties, plus all trends, need to be incorporated into these four cells to build more complete stories (scenarios) about what the future might bring for the biosciences. We use the basic scaffolding of Figure 7.2 to develop more detailed views of the future, realizing that the actual future will most likely lie in the middle. Of course, the future could evolve beyond the cone's boundaries, although we deem that unlikely. Conceptually, the cone defines a range akin to a statistical confidence interval of, say, 90% probability.

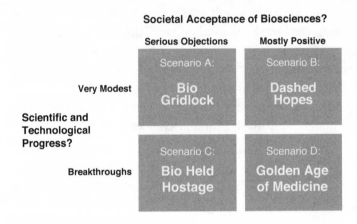

Figure 7.2   Scenarios for the biosciences

The 2×2 matrix in Figure 7.2 visualizes four possible futures for the biosciences:

1. **Biogridlock**—Molecular science has failed to live up to its high hope. This has emboldened groups opposed to stem cell research, gene therapy, cloning, and other areas of bioscience research, further dampening progress. Some of the failures and their unintended consequences, especially in gene therapy, cloning, and stem cell research, are fueling deep societal concerns about the safety, fairness, and ethical implications of the research.

2. **Dashed hopes**—Even with strong public support for science and strong public demand for biomedical solutions, technology has been slow to deliver, for technical reasons. In spite of massive research investments, the science has led to only incremental advances in healthcare. Hopes have been dashed, promises remain unfulfilled, and a sense of malaise has beset

the field, resulting in cutbacks in funding, philanthropic support, and reimbursement.

3. **Biosciences held hostage**—Significant breakthroughs have come from the biosciences, but medical side effects and ethical and privacy issues are limiting the opportunities for commercialization. Although potential has been demonstrated in the laboratory, biosciences are held hostage by negative public perceptions and cautious public policy.

4. **Golden age of medicine**—Remarkable achievements in bioscience research and development have ushered in a golden age of medicine—making it possible to cure or control most diseases, extend longevity, and improve the quality of life. Seeing the power of medicine to save lives, the public is giving unprecedented support to the biomedical community.

The role of stakeholders

The future of biomedicine is neither preordained nor entirely rooted in chance. Organizations and individuals from many quarters have the potential to influence the pace of progress. These stakeholders can impact how strong certain trends will be and which outcomes might materialize for any given wildcard. For example, if a scenario seems to emerge that violates a particular stakeholder's agenda, that group will attempt to change it. So depending on their interests, power, and ability to form coalitions, stakeholders can influence the future trajectory of the biosciences. The stakeholders in the health-care value chain include commercial companies, research laboratories, hospitals, government agencies, legislative bodies, international health organizations, insurance providers, advocacy groups, and the media. Each can influence the future of the biosciences in small as well as sometimes big ways. Figure 7.3 identifies four major clusters we need to be mindful of as we contemplate scenarios for the future of biomedicine.

In the next chapter, we explore the complex interactions among trends, wildcards, and stakeholders in greater detail, by developing the two most extreme scenarios. The actual scene in 2025 will likely

Figure 7.3   Stakeholder groups

be somewhere in between these boundary cases, which mainly serve as reference points for thinking about the future.

Endnotes

[1]Several books cover the art and science of scenario planning. For an introduction, see Paul J. H. Schoemaker, "Scenario Planning: A Tool for Strategic Thinking," *Sloan Management Review* (Winter 1995): 25–40.

[2]*The Guardian*, "Oxford University Opens Controversial Animal Research Laboratory" (18 November 2008). www.guardian.co.uk/science/2008/nov/11/animal-research-oxford-university.

[3]James M. Wilson, "Lessons Learned from the Gene Therapy Trial for Ornithine Transcarbamylase Deficiency, *Molecular Genetics and Metabolism* 96 (2009): 151–157.

8

Scenarios up to 2025

Instead of describing multiple scenarios in detail, we examine the more extreme ones, to help set the range. The real future is bound to lie somewhere in between. Using the scenario framework of the previous chapter, we focus on scenarios A and D here. They represent cases in which the two key uncertainties—societal support and technological progress—both work either against or in favor of advances in biomedicine (see Figure 8.1). We build on the wildcards discussed in Chapter 7 and explore their medical, societal, and political interconnections in this chapter. Each scenario we depict concludes with an analogy from another industry where our basic scenario story already has played out. These analogous cases suggest that our scenarios are not so farfetched, while reminding readers of their profound impacts in a different industry setting.

Bio Gridlock scenario

We start with the most negative scenario, where multiple forces undermine progress in the biosciences, and where various stakeholders conspire against advances in biomedicine, sometimes without fully realizing the consequences of their actions.

The view from 2025

The new millennium started with much hope, but as 2025 approaches, molecular science has largely failed to fulfill its huge promise. The worldwide economic problems that began to mushroom in 2007 made the rationing of medical care a reality and limited funding for medical research. This has led to apathy among many patients, practitioners, investors, and politicians, as well as active opposition from special-interest groups regarding the poor yield on resources expended and unresolved ethical issues. In this bad economic climate, groups opposed to stem cell research, gene therapy, and other sensitive areas of biomedicine are especially vocal about cutting funding for controversial research.

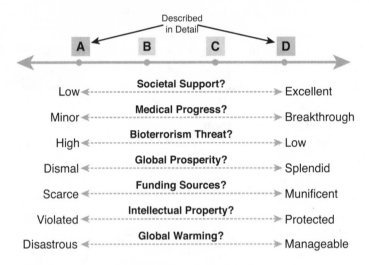

Figure 8.1   Bounding the range of scenarios

Cable television and talk radio carry far more stories on biomedical failures than successes. They highlight the risks and costs of the research, as well as egregious violations of patient privacy and other ethical breaches. Privacy issues have emerged as a major concern following the push toward electronic medical record-keeping started under the Obama Administration. The proliferation of companies that provide sequence data on part or all of a person's genome have led to media leaks on the genetic information of some high-profile individuals. In addition, medical records are routinely sold for marketing purposes. Entire regions of the world have become "anti-biosciences" zones in which much of the public opposes biomedical research for privacy, religious, political, or economic reasons.

The failure of several prominent blockbuster drugs has precipitated a serious crisis in the pharmaceutical industry. In addition to pharma companies being unable to sustain their R&D pipelines, animal rights activists (led by PETA) have created legal and public relations nightmares for these firms and for university research labs that use animals in their research or testing. The situation in the biotechnology sector is also bleak. More than half the 5,000 biotech companies that existed in 2005 no longer exist as such. Investors have lost large amounts of capital on failed biotechnology ventures. Government agencies have found it hard to justify increased funding for medical research.

How we got here

Promising technologies such as genomics, proteomics, and stem cell and gene therapies failed to deliver big breakthroughs in curing diseases. Early experimentation with therapies based on these new platforms resulted in severe adverse reactions, including death, as well as side effects that did not appear for several years after treatment. In addition, hucksters and respectable companies alike overly promoted anti-aging therapies that seemed promising, without real results. The misleading claims and adverse side effects of both legitimate drugs and "miracle elixirs" spawned class-action lawsuits, bad publicity, and more stringent regulation. The public can order thousands of genetic tests for disease genes over the Internet. However, no cures or treatments are available for the vast majority of these diseases. The failure to deliver impactful solutions have led critics to liken the bioscience industry to the early nuclear power and fuel cell sectors, where tall promises failed to materialize in a timely fashion. In addition, green technologies, material science, and nanotechnologies (non-bio) have increasingly drawn talent, money, and political support away from the biosciences.

Likewise, bioinformatics has a limited impact in healthcare, especially compared to the hype in the early part of century. Most of the focus has been on improving healthcare delivery and efficiency, including a tortuous transition to electronic medical records. The occasional use of medical records by insurance companies to deny coverage, in addition to violations of medical privacy and the use of genetic profiles in job discrimination, have resulted in many legal restrictions on who can access what kind of medical information. The net result is that the information side of medicine remains highly fragmented, with many islands of data remaining unconnected to each other. Popular media sensationalized many of the setbacks and

Highlights of the Bio Gridlock scenario

To help summarize our negative scenario, we list some of its major features. Some of these are causes of the scenario, whereas others highlight some key consequences. Cause and effect are not always clear in more complex scenarios.

- Despite years of effort, little progress has been made in bioscience research, and widely publicized adverse reactions and failures have had a chilling effect.

- Emboldened by these failures, activists have rallied opposition against the new technologies, leading to a "return to the dark ages."

- Biotech is in retreat, and pharmaceutical companies are retrenching after the withdrawal of several blockbuster drugs and vilification of the industry.

- The world economy is suffering from economic malaise fueled by geopolitical instability and continued regional military conflicts.

- With limited advances in treatment, the disproportionately large aging population is a burden on healthcare systems in the developed world.

- Attitude toward business and globalization is deeply negative.

- The prevailing sentiment and fear in society is that the biosciences have gotten too far ahead of ethics, regulation, and safeguards.

- Animal rights activists have gained much ground, making medical research with mice, rabbits, and chimps difficult and expensive in terms of compliance and PR.

- The U.S. economy has never fully recovered from the depression that began in 2008, due to high healthcare costs, a huge national debt, trade imbalances with the rest of the world, and high energy costs.

A "silent revolution"

"All the changes we've made have not done much to truly revolutionize healthcare. It remains highly vulcanized. If you go in for a surgical procedure, it is vastly inefficient. The solution and opportunity lie in being able to get more quality healthcare for much less money—but the systems to deliver this, and engagement of patients in the process, are not what they need to be to get there. The technology revolution in the next 15 years will be more of a silent revolution—people integrating a lot of stuff that is sort of lying around. The pharmaceutical industry may or may not have the money and structures to do this."

George Milne, Executive Vice President for Global R&D (retired), Pfizer

abuses, resulting in a broad public backlash against bioscience technologies and its practitioners.

Doubt and fear. Public fears of genetics-based therapies, combined with a notable shift in public sentiment in support of the moral majority and other conservative religious groups worldwide, have resulted in skepticism and rejection of new bioscience therapies. Superstitions, fears, and entrenched orthodoxies have led to a breakdown of informed and reasoned debate at the societal level. Privacy fears about genomics, moral dilemmas about genetic screening, adverse selections scandals among insurers, and distrust of large companies characterize the prevailing attitude. Furthermore, in the United States, biological sciences are further challenged in public high schools and religious circles by creationists and proponents of intelligent design whose arguments have struck a chord with the general public. The trust gap between science and society has widened.

The public rejection of bioscience technologies has taken several forms:

1. Religious pressures or concerns raised by leaders
2. Heated moral opposition in political campaigns
3. Ethical debates about abuses and setbacks in the media
4. Legislative and legal restraints, as with stem cells or abortion

5. Public support for non-governmental agencies (NGOs)

6. Refusal to approve drugs with known, limited side effects

7. Outright bans on certain kinds of technology (such as Genetically Modified Organism [GMO])

8. Withholding of government or private funding for bioscience

Critical movies such as *Dirty Little Secret* (about the black market for human organs) and *The Constant Gardener* (about drug companies committing murder to advance their profits) were harbingers of growing opposition to the industry. Subsequent real scandals about organ trading, leaking of private genomic data, and deaths in clinical trials of gene and stem cell therapies have added more fuel to the fire of public concern.

Ironically, the negative sentiments are partly fueled by the remarkable success of some technologies. For example, nanotechnology has yielded an array of devices that can measure organ performance, blood flow, and cardiovascular health. Doctors started to insert nano-sized devices into the human body to help repair organs, dispense drugs, clean arteries, and measure vital biostatistics. However, the invasive nature of these devices prompted concerns about unregulated information access and safety. Even though nanotechnology offers abundant benefits, its success fanned the fears of consumers— just as genetically modified foods did earlier. Nightmare scenarios were depicted in which genetically modified crops could spread their foreign genes to other crops nearby and far afield, with scientists helpless to contain the spread.

The potential social, legislative, ethical, and containment issues raised by nanotechnology bear much similarity to those that bedeviled GMOs a few decades earlier. Fear of proliferation of nanoparticles has gripped the public, as these minute particles reside in a plethora of medical and nonmedical products. For example, the antimicrobial killing power of silver nanoparticles has been harnessed in washing machines that release billions of silver particles. They are used to sanitize hospital laundry, and catheters are often coated with these particles to keep down infection rates in surgical patients. Household and personal care products contain nanoparticles as well, from food containers and toilet seats, to shampoos and baby diapers. Likewise, deodorants use nanoparticles, with claims reminiscent of antimicrobial soaps that appeared in the 1990s. As a consequence, nanoparticles are widely present in the environment from drinking water to ground water, with growing calls for steeper EPA regulations. Also, there is much concern that nanoparticles are penetrating the skin, being inhaled or ingested. The fear is that they accumulate in human organs such as the brain and liver, while the long-term health effects are still not known.

The global scene. "Economic malaise" best describes the state of the world economy in 2025. In the United States, years of high deficits due to wars and government bailouts of ailing industries, compounded by the entitlement demands of aging baby boomers, have resulted in high taxes and high interest rates. The global economy is in the doldrums, with frequent recessions. Healthcare costs have continued to increase, and the economy is not generating sufficient revenue to pay for full healthcare; the burden is falling again on the U.S. consumer. Low-cost generics and prescription drugs from China and India are squeezing profits from the traditional pharma companies, thus undermining the very engines of R&D that produced these products as well as new medicines.

The U.S. economy has been weakened by several crushing forces, including the huge national debt; unwieldy trade imbalances with China, India, and other rising economies; soaring energy costs; and

The impact of bioterrorism

"What happens if there is a misuse, or bioterrorist use, of bio-engineering? The worst thing that could happen is that some form of bioengineering is used in a terrorist attack—this would be the equivalent to atomic power used in warfare, and would stigmatize the industry. If someone uses bioscience in a terrorist attack or war, this literally is the worst thing that would happen. It would be very hard to move ahead, post-terror. As consumers, the public sense of 'disgust' would be high, especially if no other public benefits have appeared from bioengineering."

Arthur Caplan, Hart Professor of Bioethics, Chair of the Department of Medical Ethics, and Director of the Center for Bioethics, University of Pennsylvania

the need to address Social Security and Medicare/Medicaid costs for the aging population. Economic realities have required the NIH, NSF, and other funding agencies to cut back their grants to investigators at universities and other medical research institutions.

The world we live in

Apathy and aging. The long delay in delivery of molecular science solutions has led people in most nations to settle into a "status quo" mentality. Medical miracles exist in science fiction, but not yet in real life. Patients suffering from debilitating diseases have lost hope in the power of technology to save them. Disease advocacy groups are frustrated after decades of lobbying for research funding, given the lack of results.

Medical science is unable to deliver solutions to fatal and chronic diseases, particularly for aging populations. There is now a large disease-ridden geriatric population, placing an enormous financial burden on the economies of both developed and developing nations. Only incremental improvements have been made in the longevity of patients with long-term chronic diseases. The focus has been more on improving life quality rather than finding a cure for people with cardiovascular disease, Alzheimer's, AIDS, and diabetes.

Industry consolidation. The increasing market share of generics and copies has led to significant consolidation among pharma and biotech companies. China, India, and other Southeast Asian and Latin American countries have put much pressure on the profit margins of Western pharma companies. For the biomedical research community, the government wells have run dry. Corporate and venture capital funding is at an all-time low. Many investors were burned by the failures of the new drug ventures and have turned their attention to other areas of medicine, such as diagnostics and information systems.

Practice of medicine. Most physicians no longer have private practices. They are working in teams within larger corporate medical establishments (such as the Kaiser model). Most care focuses on relieving or stopping symptoms for patients with chronic, often life-threatening diseases. The practice of holistic medicine has increased worldwide. Attitudes toward death are changing as society can no longer afford to spend large sums in the final phase of life. Passive euthanasia, hospice, and assisted suicide are on the rise. Books such as *Dying Naturally* and *Misplaced Medical Heroics* make the bestsellers list.

The traditional doctor/patient model still exists, but significant changes have occurred. Telemedicine allows for more in-home care; more clinics exist to deal with the burgeoning number of Alzheimer's patients; and government-funded care of the elderly has expanded, as in Programs of All-Inclusive Care for the Elderly (PACE). The shortage of quality healthcare personnel, especially for geriatric care, has become even more acute. Healthcare workers from less-developed nations continue to be in demand to fill the gaps. But since the quality of life in emerging countries has improved, foreign medical students and doctors who attended U.S. universities often return to their home countries. Also, fewer foreign students are coming to the United States to study since other nations have strengthened their own educational systems.

Consumer representation. Groups that can speak for the consumer are exerting increased pressure on legislative and regulatory bodies about the public funding of research and treatment. Because people over 55 require the greatest amount of care and are now the largest cohort, their voice has become the loudest. Representative groups such as AARP, AAHSA/CAST, and the Alzheimer's Association are a major force. Nonetheless, there is a widespread feeling that much of their lobbying effort to bring about better, cheaper healthcare has been largely ineffectual.

Analogous cases

Some may doubt that bio gridlock is a plausible scenario. In the spirit of scenario planning, the future depicted here is a more extreme version of what we think could happen. Indeed, the essence of this scenario has already played out in other industries, notably nuclear power and genetically modified foods.

Nuclear power. On March 28, 1979, Reactor 2 at the Three Mile Island (TMI) nuclear power plant in Middletown, Pennsylvania, suffered a partial meltdown, resulting in more than 2,000 personal injury claims due to gamma radiation exposure and more than $50 million in payments to plaintiffs. Radiation readings were far below "health limits." Nonetheless, this accident initiated a long, slow downward spiral for the U.S. nuclear power industry. No reactors have been built in the United States since 1979, 74 reactors under construction were

Sample headlines for Bio Gridlock

To help visualize this scenario, we offer the following illustrative headlines that might appear in future news bulletins. These are not meant to be exact predictions but rather illustrative examples of events that are consistent with the basic theme of this scenario as it plays out over time. Some of these events could happen earlier, later, or not at all. And other, even more striking ones, might make the headlines instead.

- 2010: Five Major Drugs Recalled—FDA Tightens Reins Further
- 2011: Wal-Mart Announces New Diagnostic Imaging Centers in 500 Stores
- 2012: Record Number of European Patients Visit India for Medical Treatment
- 2013: New, Extremely Virulent AIDS-like Disease Emerges in Africa; AIDS Drugs Have No Effect
- 2014: Many African Countries Lose Battle Against Mutating HIV-AIDS
- 2015: U.S. Budget Deficit Exceeds $10 Trillion, Income Tax at 65% Max
- 2016: Predictive Diagnostics Tell Patients Future Disease Profiles
- 2017: Asian Doctors Attempt the First Full Human Brain Transplant in an Alzheimer's Patient
- 2018: Mothers Taking New DNA Vaccines Have Higher Birth Defects
- 2019: U.S. Enacts Truth in Research Legislation, with Criminal Penalties
- 2020: First Global Guidelines for the Safe Use and Disposal of Products Containing Nanoparticles
- 2021: European Union Charges Private Nursing Homes with Gross Negligence

- 2022: Million Person March in Washington to Protest Big Bioscience Business
- 2023: Human Cloning Reported in Clandestine Asian Hospital for the Rich
- 2024: Black Markets for Human Organs Flourish in Poor Countries
- 2025: Strong Animal Rights Legislation Enacted; PETA has Pharma in a Corner

cancelled, and 13 plants operating at the time were shut down. The Chernobyl nuclear meltdown in 1986 in the Ukraine further inflamed fears about the technology's safety systems. Antinuclear protestor groups gained prominence in the United States and elsewhere.

Genetically modified organisms (GMOs). In 1994, Calgene introduced the genetically modified "Flavr-Savr" tomato to the U.S. market, and later Monsanto introduced corn and other crops that are resistant to particular herbicides. Since those pioneering launches, opposition to genetically modified organisms (GMOs) has spread across international boundaries, political ideologies, religious groups, and activist causes. GMOs have been banned in several countries, and in many others, foods must be labeled if they contain genetically modified ingredients or processes. Those opposed to GMOs voice four arguments:

1. Genes from GMOs can spread to other crops or weeds, causing harm.
2. Corporate entities do not have the right to modify plants or animals and to hold patents on these life forms.
3. Foods from genetically modified plants or animals might pose unknown health risks for humans who consume them.
4. Social and economic consequences, such as the impact on small farmers, could be detrimental and unfair.

Golden age scenario

We now turn to the other side of the scenario spectrum and paint a very positive view of possible developments in biomedicine. As before, the aim is not to make precise predictions but bound the range of possible outcomes in terms of the most optimistic view. The actual future in 2025 will likely be in between the two boundary scenarios depicted in this chapter.

The view from 2025

Remarkable achievements in molecular science and other areas of medicine have ushered in a golden age of medicine. Biomedical solutions are making it possible to cure or control most diseases, extend longevity, and improve the quality of life. Previously fatal conditions, such as cancer, cardiovascular diseases, and Alzheimer's disease, are increasingly being conquered. People are living longer, healthier, and more productive lives, thanks to bioscience technologies.

Breakthroughs in regenerative medicine have extended the life expectancy of people in developed countries, exacerbating the demographic and financial pressures of an aging population in the United States and other nations. Millions of people in the developing world have benefited from novel treatments for malaria, AIDS, TB, and many other diseases that have plagued the third world for centuries.

Seeing the power of medicine to save lives, the public has given unprecedented support to the biomedical community. Informed by the educational system and the media, the public fully embraces the new scientific and technological advances. Investment funding from both public and private sectors has flowed like a swollen river into bioscience research. Biotech firms with life science products have grown larger than the largest pharmaceutical firms, shaking up the industry.

How we got here

Breakthroughs in science. Vast searchable libraries of genes and proteins have enabled scientists to use molecular information to identify segments of the population that will benefit most from new types of drugs, while restricting use by patients for whom the drug will be harmful or ineffective (personalized medicine). Advances in stem cell research have yielded many new therapies, such as brain cell replacement in Alzheimer's patients, a cure for Parkinson's, and the ability to grow organs. Tissue engineering, organ replacement, and regenerative medicine have greatly improved the length and quality of life for senior

Scenario highlights

Here are the essential elements of this scenario, representing a mix of causes and consequences:

- Breakthroughs in biosciences have led to new cures for many chronic diseases, as well as an expansion of personalized medicine.

- Encouraged by these successes, public support for biomedical research and treatment has reached record highs.

- Healthcare consumers feel empowered by advanced diagnostics and information technology.

- Global economic growth remains strong with minor armed conflicts.

- Biotech firms and new entrants compete strongly with the traditional pharmaceutical companies.

- Mental illness treatment and compliance are vastly improved; prison populations shrink.

- Medicine and healthcare are personalized as well as global.

- The biological clock is yielding its secrets and slowing down aging.

citizens. Science is working aggressively on ways to slow or reverse the aging process.

Discoveries in molecular medicine have led to novel forms of treatment for many diseases that have plagued underdeveloped regions of the world for centuries. New flu viruses and drug-resistant bacteria continue to emerge, but international surveillance allows for rapid detection and genotyping of new strains. Also, vaccine production times have been cut dramatically and novel antibiotics have largely conquered resistant bacteria, which were once a scourge of hospitals.

Public support is soaring. Strong public support for new biomedical technologies circulates around the world. People appreciate the capability of biomedicine to turn fatal diseases into chronic and, increasingly, curable diseases. University hospitals and research centers recorded more contributions in 2023 than any other year, thanks to corporate and individual benefactors expressing their gratitude for medical cures that saved the lives of family members or employees.

Progress in genomics and proteomics

"The payoff for progress in genomics and proteomics is enormous. These developments will lead to individualized medicine, treatments based on genotypes versus phenotypes, and higher success rates due to more tailored drugs and delivery. Drug delivery will also benefit from technological advances. Nanotechnology will be used to assemble molecules and target areas in the body. Serious diseases may benefit from delivering genetic drugs directly to affected organs."

Andrew I. Schafer, MD, Frank Wiser Thomas Professor & Chair of the Department of Medicine, University of Pennsylvania Medical School

The surge in success and support for new medical technologies has provided unprecedented funding for medical research and relaxed regulations. This has allowed many treatments to be commercialized and licensed faster than in the past. The public also recognizes the capacity of technology to improve the economics of healthcare. This is especially true where generic drugs and international pricing pressures have lowered drug costs. Increasingly, patients from developing nations such as China and India are able to obtain more sophisticated medical treatments.

Diagnostics and information. Confidential diagnostic testing is available to individuals as well as hospitals, clinics, and physicians. Most tests are available over the Internet or by submitting mail-order samples to private laboratories. Consumers can purchase home tests for most diseases at drugstore chains and supermarkets. A handful of diagnostic labs have grown to multibillion-dollar global franchise operations, delivering lab results directly to consumers by email.

Medical information technology continues to empower consumers. People in most countries have access to online medical information, including the same services available to medical practitioners. However, the surge in online medical communication has attracted medical quacks and misinformation from many sources. This has required the creation of special enforcement agencies called "medical web police" whose responsibilities include eliminating online medical disinformation.

Peer-to-peer structures allay privacy concerns

"The notion of a massive database is interesting, but it's the individual who needs to take control of the data. Individuals will be very leery about sending over data and not knowing what will happen to it. One of the solutions is a self-organizing peer structure around certain symptoms. If you look closer, that already happens. Public–private partnerships (PPPs) support development of drugs, vaccines, and diagnostics to address diseases that predominantly afflict the poor, such as HIV/AIDS, tuberculosis and malaria. There is one orphan disease where people got together and raised $500 million for universities to study it. Otherwise, pharma would not have worked on this."

Ulrich Goluke, Manager, Scenario Unit, World Business Council for Sustainable Development (WBCSD), Switzerland

The global scene. The strong global economy is an ally. High-end pharmaceutical development stays within the United States, while lower-end development is outsourced. Success has also created economic strain. Many cures and treatments exist that Medicare, Medicaid, Social Security, and other public and private insurers just cannot afford to cover. The nursing home "investment bubble" burst in 2015, when fewer geriatric patients required nursing care and could remain at home in relatively good health. A wide variety of products to address age-related disorders are available from pharmaceutical, biotech, nutraceutical, diagnostic, cosmetic, and food companies. Biomedical solutions in the field of regenerative medicine, pioneered in Korea and Japan, have helped to extend the longevity of retired persons, straining the pension, Medicare, and Social Security systems. However, because many senior citizens remain in good health, they continue to work well beyond retirement age, which has increased to age 70.

Bioterrorism. The threat of biowarfare has remained high as impoverished and unstable nations develop significant capabilities in chemical and biological weapons. Wide sharing of research information via the Internet has put advanced technologies for genetically manipulating and dispersing deadly pathogens into the hands of unstable dictators, rogue nations, and terrorists. On the other hand, biometrics, sensors, chemical detectors, and other technologies kept pace such that richer nations developed a "bioshield" that protects them against attack.

The world we live in

The rise of biotech. The largest global provider of bioscience products is a marketing organization created and sustained by a consortium of biotech firms that have pooled their technologies and products to compete with large pharmaceutical giants. They form a "bio-*keiretsu*" (referring to the Japanese model of highly interconnected suppliers and wholesalers).

Much of the success in new life science technologies has come from biotech ventures, university research programs, and research centers sponsored by governments and corporations. Pharmaceutical firms continue to acquire biotech firms so that they can incorporate their discoveries and pipelines into their own organizations. Many biotechs accepted investment from foreign nonhealthcare firms, especially in the energy and IT sectors, and pose strong competition for the few remaining "Big Pharma" firms. Indeed, some biotech firms have acquired old pharma companies to get closer to the market.

Patient monitoring. Real-time monitoring has become a major part of medical practice. Healthcare providers are notified when patients skip pills or take the wrong medications. In many cases, they can monitor patients 24×7 through wireless implants.

The growth of genetically engineered tissues and organs in "tissue farms" has taken personalized medicine to a new level, extending the lives of many people who would have otherwise died of "natural causes" such as organ failure. Artificial organs and medical implants have provided alternative choices for those who are opposed to tissue farming on religious, moral, or safety grounds.

The "killer app" in the field of informatics employs a combination of nanosensors, semiconductor implants, wireless communication, and other technologies to monitor patient health. This cuts down on the number of doctor visits or days in the hospital. Those at risk of heart attack can be monitored using tiny sensors that are the size of a grain of rice, with a battery life of several years.

Total healthcare. The success of biomedical technologies in preventing, treating, and curing disease—and reducing healthcare costs—has resulted in many developed countries adopting the Total Healthcare Management (THM) model. Forms of this model are also being offered in China, India, and other developing nations with large populations. Preventive medicine has improved dramatically, thanks to diagnostic tests, new forms of scanners that expose patients to far less radiation, and intrabody sensors that provide new patient-monitoring capabilities. Advances in genetic engineering have eliminated many single-gene defects by intervening before birth. Nutraceuticals and genetically engineered foods contain vaccines and drugs that can prevent many of the dreaded diseases of the developing world.

Practice of medicine. Medicine focuses increasingly on prevention. The medical field is enriched by bioengineers who oversee automated diagnostics using sophisticated computer models of the patient. When a person's genetic propensity for certain diseases is known, experts can recommend appropriate diet and lifestyle choices.

Because of the advances in diagnostic tools, nurses and technicians do most of the diagnostic work and dispense most treatments. Doctors focus on designing treatment plans for the patient. However, some specialties, such as plastic surgery and artificial organ transplantation, still require skilled physicians and surgeons. A growing percentage of

Outpatient care for heart attacks

"If technology delivers on its promises, I could see a fundamental paradigm shift from treating diseases to curing diseases that would enable people to live longer. I don't necessarily mean an 'absolute' cure, but maybe a partial cure requiring periodic booster shots or ongoing diagnostics to make sure that either the condition doesn't recur or side effects are curbed. For example, picture a patient who had just suffered a heart attack. If, instead of receiving a heart transplant, the patient's heart could be injected with regenerative stem cells, it would have an enormous impact on hospitals, especially if the procedure could be done on an outpatient basis."

Thomas Tillett, President, CEO, and cofounder, RheoGene

surgeries can now be performed remotely by robotic arms, with error rates of .00001% or less. U.S. and European doctors are also facing competition from new "global healthcare centers" in India, China, and other countries. Patients from around the world fly to these centers to gain access to sophisticated technologies that cost much less there than in their own countries. Often insurers reimburse their expenses, including travel, given the lower costs of these offshore treatments.

Mentally ill patients who require ongoing medication are now monitored remotely by their healthcare providers. Surgically implanted biosensors keep track of blood levels of psychiatric drugs, and nanoscale drug delivery particles keep medicine at specified levels in the bloodstream for those patients unable to manage their own medications. This has resulted in fewer cases of recidivism for criminals who are mentally ill, as well as a significant drop in legal and judicial costs for repeated prosecution. Another benefit has been a sizable drop in the prison inmate population.

Consumer behavior. Consumers are greatly empowered. Diagnostic advances and consumer access to online information lead patients to diagnose themselves for many common ailments. In simple cases, they put together their own treatment plans, limiting the need for doctors and reducing the burden on the healthcare system. Patients welcome social support networks, such as vlogs, blogs, wikis, and online communities dedicated to specific medical problems and solutions. Healthcare providers and insurers are walking a tightrope on privacy, with the vast expansion of genomic data on individuals. HIPPA is expanded to cover other potential misuses of patient and consumer data.

Analogous case: information technology

To some, the golden age of medicine may seem too good to be true. However, this kind of positive scenario, driven by technological advances as well as key market developments, already played out in the semiconductor industry. A "new age" of electronics was born when the first generation of microprocessors were invented in the 1970s. Converging technologies allowed a small microprocessor to be imprinted on a single chip, fueling the growth of video games and personal computers. In the mid 1970s, the Intel 8080, Motorola 6800, and MOS

Sample headlines: golden age of medicine

As with the negative scenario, we list here some illustrative headlines that might appear if the golden age of medicine becomes a reality. They are only meant as examples, not exact predictions in terms of timing or details. The underlying ideas may also appear in different guises than suggested here.

- 2010: Ban on Embryonic Stem Cell Research Lifted
- 2011: Universal Healthcare Fully Enacted in USA
- 2012: End of the Blockbuster Drug: Generics Win
- 2013: Vaccination of Boys for Cervical Cancer is Common
- 2014: Singapore Magnet for Bioscience Research
- 2015: Nobel Prize in Medicine Goes to Stem Cell Research
- 2016: Biochips Routinely Implanted in Patients
- 2017: Breast Cancer Treated Only with Pills Now
- 2018: First Repair of Genetic Defect in Utero
- 2019: Diabetes is Largely Cured, Thanks to Stem Cell Therapy
- 2020: Heart Transplant Performed Remotely by Robotic Arms
- 2021: Tissue-Farming Centers Established in China and India
- 2022: New Treatments for Aging Disorders Extend Human Lifespan
- 2023: Reproductive Cloning Embraced for Couples Who Lose a Child
- 2024: Many Blind People Can See Again: Converging Technologies
- 2025: Census Report: Over One Million Centenarians Alive in U.S.

Technology 6502A chips enabled the development of the first Bowmar, HP, and TI handheld digital calculators, as well as the first personal computers from Apple, Radio Shack, and Commodore. Personal computers and cell phones have become the largest use for semiconductors.

The stunning success and demand for these devices drove hand-held calculator prices from their early 1970s price range of $300 to as low as $49 by 1975. Personal computer prices fell from more than $1,500 to $300 by 1980, and digital watch prices fell from $100 in 1977 to less than $10 in 1979. Apple was founded in 1974, and Microsoft in 1977. In 1983, Motorola's first cellular phone was brought to market and the creation of national cellular infrastructure networks in the mid-1990s created a revolution in mobile—and global—telecommunications. Multiple technological breakthroughs, from the transistor, to a microprocessor put on a single chip, to advances in wireless and Internet search, allowed entirely new industries to emerge, with such major global companies as Apple, Microsoft, Cisco, Google, Nokia, Comcast, and numerous telecommunication giants. Similar innovation, wealth creation, and new giant companies are likely to emerge in the golden age of medicine over the next few decades.

9

What it all means

In this closing chapter, we take a pragmatic look at the healthcare implications of the biosciences for you, your family, at work and for society as a whole. The specific impacts will depend on which scenarios occur from the wide range of possibilities. But the scientific and medical march of the biosciences is beyond dispute. The only uncertainties are when and where they will impact us most. Readers of this book are armed with the knowledge and conceptual frameworks necessary to follow new developments as they play out across the globe. The following sections are organized by areas of impact on your family, your career, commerce and industry, and society at large.

You and your family

Medical diagnostics will see new advances sooner than therapeutic interventions due to differences in research, development, and regulatory cycles. Diagnostic tests will become ubiquitous (from the internet to shopping malls and airports), inexpensive, and easy to conduct, often without medical oversight. This abundance of diagnostic measures will present vexing problems for many of us. How much do we really want to know about our health, especially when it is unclear how much we can alter it? In addition, questions will arise about the reliability of the tests, interpretation of results, and what to do with them. As with current amniocentesis tests for the unborn, or full-body CT and MRI scans for adults, future diagnostic procedures might reveal abnormalities that are harmless. This leads to ambiguous implications for treatment, or unnecessary concerns about potential medical conditions for which no treatment yet exists. Rational decision theory suggests that we should agree to diagnostic tests only if we are willing to act, given the available treatment options, on at least some of the results that might emerge. For example, if a couple expecting their first baby is not willing to act on any of the possible outcomes of an amniocentesis test, such as Down's syndrome or spina bifida, then why do the test? But humans are not fully rational and might not be able to answer the hypothetical "What if?" questions unless the diagnosis is actually rendered and the medical decision situation is real. Just consider how hard it is to construct a living will when still fully healthy and far from death, and how our opinions might shift over time.

Information about our genetic predisposition for disease is going to increase dramatically in the coming years, and much faster than medicine's ability to treat these diseases. Today parents can have their embryos screened for diseases such as Tay-Sachs and cystic fibrosis. In the near future, parents might be able to select embryos with specific eye color, athletic ability, intelligence, or perhaps even complex personality traits. Even further down the line, embryos might be cured of presumed defects. Questions will arise about whether embryos should be screened in utero for violent tendencies or for destructive sexual deviance such as pedophilia or sadism if scientists know the genes involved. More controversial yet, should parents be allowed to select their child's gender or perhaps influence sexual orientation, if genes for homosexuality are found that scientists could alter with gene therapy? Few of us are ready for this brave new world of daunting choices. Nonetheless, clear benefits will exist in the case of preventable genetic diseases and other maladies. For example, in the future, families will have more options for dealing with infertility (see sidebar) and for enhancing the traits of their offspring.

In addition to these more futuristic possibilities, we need to exercise more care today about what medical records to share with others. Apart from privacy concerns, there is the risk such information—much of which is digitized rather than stuffed in manila folders—will be used against us. For example, insurance companies might promise discounts for life or health insurance if people consent to additional tests (beyond a routine physical exam, EKG, blood test, and personal medical history). However, this might be a Faustian bargain if it results in higher premiums or exclusion of preexisting conditions. Medical records might be shared without people's knowledge and

Advances in fertility treatment

Since the birth of Louise Brown, the first in vitro fertilization (IVF) baby, in 1978, the field of fertility treatment has made huge progress. As more parents delay their first pregnancy, couples who could not previously have children can now often be helped.

Major fertility advances of recent decades include the following:

- **In vitro fertilization (test-tube babies)**—Originally, IVF was used to treat the 15% of couples whose infertility was due to blocked fallopian tubes. Today, it is used in cases where sperm count is low or when the woman's immune system inhibits sperm motility.

- **Intra-cytoplasmic sperm injection (ICSI)**—Injecting sperm cells directly into the egg is routinely done today under a microscope. This especially helps couples when the sperm count is low, if sperms are not moving well or if egg and sperm fail to fuse.

- **Recombinant DNA technology**—This process creates purer hormonal medicines used to stimulate the ovaries for egg production. These replace urinary preparations that can cause allergic reactions.

- **Freezing of embryos and unfertilized eggs**—Left over embryos can now be safely frozen and then thawed for future implantation attempts. This can also help women whose cancer treatments have destroyed their fertility, by implanting their previously frozen, unfertilized eggs.

- **Preimplantation genetic diagnosis (PGD)**—Doctors can screen out embryos with defects before transferring them to the mother. Such screening has been done for diseases caused by a single gene defect such as Tay-Sachs or Huntington disease.

Exciting future possibilities in the fertility field include the following:

- **PGD using DNA chips**—Genetic profiles of all embryos may be developed, to screen for many types of disease as well as for other traits such as intelligence, athletic ability etc.

- **Combining hormones**—By combining hormones, they can be made to stay in the body longer, allowing weekly injections to stimulate the production of eggs rather than daily ones.

- **Ingestible hormones**—These would be more stable in the gut and could pass through the intestinal wall and into the bloodstream, with the same effect as current hormone injections,

- **Helping cancer patients**—To better preserve fertility, doctors could remove and freeze tissue from ovaries or testicles. A few years later, the tissue could be thawed and stimulated with hormones in the laboratory to produce eggs or sperm cells.

- **Genetic engineering of embryos**—Using biotech tools, we will be able to change the genetic destiny of our children by replacing bad genes with good ones or simply adding new, positive traits to the embryo.

- **Cloning**—Reproductive cloning of humans is only a matter of time, in order to replace a dead child or allow infertile or homosexual couples to have a child without using genes of a third person.

Based on contributions by Hans M. Vemer, M.D., former President of Organon International, now with Schering-Plough Corp.

could come back to haunt them in surprising ways, such as when applying for a new job or proposing marriage. Such adverse selection, which is already occurring today in insurance markets with respect to medical preconditions, obesity, and smoking, will get stronger with additional information. If unfettered, it will lead to fully personalized risk pricing, in which each individual is assessed a premium for health or life insurance that reflects that person's unique genetic profile and life history. Apart from undermining the risk pooling basis of insurance, it could lead to cherry picking among insurers and exclusion of those most in need of medical coverage. Legislation is being drafted in the United States and elsewhere to guard against this.

Another future challenge posed by advances in biomedicine is how to manage the dynamics of family structures in which six or more generations are alive all at once. Apart from the need for better financial planning and different estate arrangements, these new family structures might affect marriage and kin relationships. Will serial monogamy become the norm if being married for 80 years to the same person isn't appealing to everyone? Should the elderly share their wealth earlier with relatives who offer support structures? Can the "sandwich" generation, who cares simultaneously for its own children and parents, also handle grandchildren and grandparents? In addition to the increased family size and the burdens of degenerative diseases among elderly relatives, the problem of genetic distance arises. According to Dawkin's selfish gene theory, our affinity for offspring declines geometrically with each generation in proportion to the percentage of our own DNA present in our predecessors or progeny.[1] If true, your great-great-grandchildren will be quite distant genetically (sharing just 6% of your genome) and new family structures will likely evolve, with weaker ties between generations farther apart in time.

An important implication of increased longevity is that medicine will need to reorient itself toward chronic care instead of acute disease, with a much greater focus on prevention and wellness. Suppose an elderly man is brought to the emergency room after a fall down the stairs in his apartment building. Traditional medicine would stabilize the patient, try to heal bone fractures, and alleviate pain. A broader medical approach would ask why the person fell and delve into personal issues. Healthcare providers might find that the man lives alone, has become an alcoholic, and takes his medicine irregularly. Furthermore, he receives few visits from friends or relatives. He shows symptoms of depression and early dementia. What is the medical problem here: physical impairment, mental issues, social isolation, and/or limited financial means? The answer is all of the above, even though medical students today are primarily trained to address a subset of the interacting variables. This scenario would require more holistic care involving a wide range of professionals, such as social workers, gerontologists, nurses, physical therapists, and perhaps ethicists or spiritual guides.

For people with above-average financial means, new options will become available to access overseas medical expertise, which is often faster and cheaper. The appeal of medical tourism will expand far beyond where it is today, conjuring images of nose jobs or facelifts in Brazil or Mexico. Also, telemedicine involving remote diagnosis or robotic surgery will be common place making it possible for more people to access the global experts in any given specialty. A wider array of medical choices will be available at home and abroad, including the good, the bad, and the ugly. The *good* will include better options for cancer, which will be treated no longer with sledge hammers (such as surgery, chemotherapy, and radiation), but with highly customized drug treatments that target tumors with great precision (as monoclonal antibodies already do today). The *bad* will include dubious treatments in cheap overseas clinics that prey on desperate patients, and domestic treatments that feed false hopes about diets, beauty, and longevity. And the *ugly* might involve criminal theft of kidneys (as in the movie *Dirty Pretty Things*), organ trades, and abuse of uneducated poor subjects from the developing world in clinical trials. Hopefully, tissue regeneration utilizing stem cells will reduce the need for illegal organ trades.

At work

New biosciences careers will be available to many workers along the full value chain of healthcare, especially to young people now still in school. Entirely new bioscience companies will emerge, similar in size and growth to Apple, Microsoft, Cisco, and Google, thanks to the confluence of information technology and life sciences. Major shakeouts will happen as well, as we saw in information technology, where thousands of computer, dotcom, and telecom companies tried and failed. Early movers don't necessarily win, and often only a few major new players survive. For example, in the biotech industry, more than 2,000 companies were founded since 1970. Only a handful remain today as independent biopharmaceutical enterprises (such as Amgen). The others didn't all fail; large pharmaceutical companies acquired many of them. High return and high risk will characterize the job opportunities in biomedicine and related fields, benefiting those who are knowledgeable, flexible, and able to manage risk.

Those already working in healthcare and related fields will face a pressing need for continuing education. Bioscience advances might affect their jobs directly, in terms of how they actually perform their work, or indirectly, perhaps by changing the nature of competition and the industrial segments that grow versus shrink. For example, one of us worked as a bench scientist in the 1980s and was amazed by how much lab instruments and procedures improved in just two decades. Considering the relentless march of technology, the key is to focus education on lifelong learning and flexible career paths, in the expectation of more frequent career shifts. As people live longer, we will see even more generations working side by side, from the very young to perhaps centenarians. This will require greater cultural and intergenerational sensitivity to get the best from the very diverse work teams we shall be operating in.[2]

Biomedical advances might also impact the conditions under which we work, or the ways in which we can improve productivity. For example, advances in light-emitting diodes (LEDs) will greatly improve the quality of illumination at work. Apart from better visibility and less eyestrain, indirect health benefits might also arise. For example, premature infants in neonatal wards tend to do develop better under LED lighting than with regular illumination. Bioscience technologies might more quickly detect "sick building syndrome," or those at risk of multiple chemical sensitivities due to noxious exposures at home, on the job, or in public spaces. Continuous monitoring of air quality in buildings, trains, and airplanes will help to detect and remove pathogens before they spread. More generally, biomedicine might improve our approach to fatigue prevention, mental concentration, sensory overload, carpal tunnel risks, or other maladies related to sedentary or stressed lifestyles. For truck driving jobs or those that require travel across time zones, new and effective regimens will emerge to combat fatigue and jet lag, customized to the person involved.

Finally, thinking strategically about what healthcare insurance policies to sign up for will become more important. Assuming that most policies will still be offered in the future through employment, it is critical to understand what is covered and for how long after an employee leaves the company. As we work and live longer, insurance for disability and long-term care becomes increasingly important, because each of these expenses can deplete a family's savings. As we choose among job offers and companies, we need to consider what they offer in terms of healthcare protection and benefits. Some employers will reward healthy lifestyles, such as controlling weight, exercising, and being drug-free since it might lower their group insurance rates. Thus, access and ownership of healthcare information becomes important, as does privacy protection, since more medical information is being collected and stored about us. In occupations in which security is important, we can expect more invasive biometrics, from iris scans to DNA markers. New bioscience technologies will be used for routine personal identification and for high-level security clearance, including perhaps the implantation of biochips with crucial health- and job-related information. In sum, we can run but we cannot hide. Employers will know a lot about us—probably more than we want.

Business and commerce

Bioscience-based innovation will pose significant challenges to industry and commerce, as new technologies have throughout business history. We can learn several relevant lessons from the past. First, commercializing unproven technologies is an extremely difficult task at which few companies succeed repeatedly. But the rewards can be great, and those that successfully commercialize bioscience technologies can easily overtake their established competitors in the market.[3] Over time, industry leaders tend to become losers because they fail to recognize new innovators and do not respond in time to prevent their own demise. The strategy literature offers various prescriptions for this danger, emphasizing that an external orientation and ability to change are vital in ensuring future viability.[4]

In our own research at Wharton, we examined extensively why established firms tend to fall into four common traps when facing new technologies.[5] These historical pitfalls are all highly relevant to companies competing in the biosciences. The first trap is that established companies are slow to invest in new technologies, because of conflicting opinions, divergent interests, and the inability to foresee future possibilities. Second, in an effort to remain in their comfort zone and familiar niche, established players tend to pursue only familiar technologies. With any new technology, leaders may hesitate to strongly commit to change, which is the third trap. However, a reluctance to commit fully can result in missed opportunities. Finally, company leadership needs to continually push the process of innovation. Lack of this kind of persistence is the fourth trap we identified for incumbent players in a variety of industries going back many decades.

Our research suggests that managing emerging technologies, as opposed to the core business, is truly a different game. Companies cannot approach the new opportunities that the biosciences present with the same techniques, mindsets, and organizational structures that support the mainstream business within established divisions. Instead, managing innovation entails paradoxical challenges. For example, even though a strong commitment is necessary, an organization must keep its options open. Microsoft illustrated this by writing software for Apple, UNIX, and IBM during the 1980s, thus keeping its options open as users made choices about operating systems. In many other instances, companies that succeed in the long run commit early to a new technology. Yet most pioneers in innovation fail. Managers must learn how to balance commitment and flexibility to thrive in multiple future scenarios.[6]

Another paradox is that innovation strategies should build on a company's core competencies (the stick-to-your-knitting mantra), yet success often requires organizational separation. In the 1980s, Motorola had lost its lead in the stereo and computer chip businesses, was unrecognized in the television industry, and was even starting to miss opportunities in the cellular market. But CEO George Fisher engineered one of the most remarkable rebounds in corporate history by clearly separating Motorola's new businesses from the old. Fisher's talent for managing new technologies shone again when he joined Eastman Kodak in 1994 to help revitalize the company's struggling photo business. Fisher separated the old chemical emulsion businesses from the digital ones and enabled digital imaging technology to flourish. Biomedical companies, especially pharmaceutical firms, will have to determine carefully what the optimal degree of separation is between their old business and new bioscience initiatives.[7]

A final paradox concerns the conflicting notions that industry success usually involves intense competition, yet ultimately depends greatly on collaboration. Companies must balance collaborative and competitive relationships with other industry innovators (known as co-opetition).[8] The pharmaceutical industry is an example of where rivalry is especially intense with generics, yet some large drug companies such as GlaxoSmithKline have teamed up with generic manufacturers to launch "authorized" generic drugs to keep others at bay.

Although no simple recipes for managing emerging technologies exist, several approaches can lead companies in the right direction. First, successful management requires wider peripheral vision.[9] It's easy for firms to become so focused on specific goals and numbers that they lose sight of the broader activities in their industry and beyond. Innovation tends to occur at the fringes, so narrowly focusing on the traditional business will cause companies to miss crucial new developments. Second, innovation can succeed only in an organizational culture that values learning and permits some degree of failure. However, the focus of most companies is on short-term performance and growth, and avoiding mistakes. Third, management should provide some degree of organizational autonomy, such as placing emerging technology initiatives in separate units, away from the parent organization. This allows the new endeavor to focus on its own value creation without jeopardizing the core business or being constrained by its rules and procedures (as illustrated with the Kodak example). Nor will it be saddled with its overhead.

Business leaders will likely face the following challenges in the biosciences: emergence of new business segments and entire industries (as we saw in the computer field); continued globalization, including the emergence of Asian partners and competitors; more complex regulations about privacy and data sharing; higher ethical expectations, with more industry regulation; a restructuring in the practice of medicine, with healthcare becoming more patient oriented and better integrated; continued convergence of IT and life science technologies; greater complexities in developing and protecting intellectual property; the need to move away from firm-centric views of strategy to network-oriented perspectives; and new opportunities to apply bioscience solutions to large-scale problems such as global warming, energy needs, clean water shortages, deforestation, and epidemics.

Society at large

Eminent economist Kenneth Boulding published an interesting social thought experiment in the 1960s.[10] He wondered how our world would change, economically and otherwise, if people lived until age 200. He started by noting how dependent our social institutions and structures are on the traditional life span and life phases, from family to school to work.

In purely economic terms, longevity will decrease people's chances for promotion (due to crowding at the top of the pyramid), profoundly affect savings and investment plans, and potentially disrupting the production of economic wealth. Normally, the young and the old consume more than they produce, whereas the middle of the age pyramid supports the rest. If these proportions fall out of balance, economic chaos might ensue, although much depends on how longevity comes about. Suppose death took a holiday for more than 100 years, such that all death would be postponed by a century: What would be the effect? Apart from putting morticians out of business and pushing our entitlement programs into insolvency, other major economic and social dislocations would occur. In contrast, if life expectancy would gradually increase, we could adjust our contracts, pension plans, educational system, legal contracts, and the institution of marriage, although still the issues of power concentration and ossification would remain.

The old will not likely surrender their privileged positions at the top, especially if their mental health remains strong. For example, we would have a surplus of old doctors, aged lawyers, tenured professors, gray-haired CEOs, and old-fashioned politicians, whose increasingly outdated ideas would become formidable obstacles to innovation and change. Starting with a clean slate of knowledge, assets, and values every 70 years is a societal virtue. It replenishes the system with new blood and refreshes the stock of human knowledge. Just as important, death reduces the amount of unlearning and forgetting that is necessary for society to function in a changed world. Perhaps only after longevity has been greatly stretched will society fully appreciate that death is a benefit to humankind as well as a curse.

With prolonged life span, the inverted age pyramid will spark profound tensions about who is entitled to pensions and healthcare, and who pays for them. Intergenerational subsidies will be unavoidable and political movements will rise to stem the inequity. The field of elder law will get a big boost, both to protect senior citizens from abuse and to safeguard their social benefits. New laws will be enacted to better protect the old, especially when suffering from dementia or are otherwise vulnerable. Political leaders will start to rethink the wisdom of trying to extend life, because of its dislocating effects on society—from a huge economic burden to rampant age discrimination against the elderly. Political power will shift toward the massive elderly voting segment, who will become increasingly demanding. An enormous need and opportunity to reeducate old healthcare workers and train new ones, who might be immigrants from developing nations, will occur amid a flourishing of geriatric subspecialties.

All this will raise new moral issues, as the field of ethics feverishly tries to catch up with technological advances in genetically repairing or enhancing the embryo, cloning animals and humans, and creating artificial life forms.[11] Moral questions will abound around the beginning of life (from abortion to gene therapy or cloning of embryos) as well as the end of life (such as artificial life support, living wills, euthanasia, and the right to die). More resources and attention will be devoted to mental illness and degenerative diseases as we age. A greater need for protection against bioscience abuses, from terrorism and human cloning to invasion of privacy and adverse selection in job search and insurance coverage, will arise.

Eventually, as we really learn how to change the genetic makeup of our offspring or introduce artificial life forms that are very human-like, we will enter a post-human era. When show dogs or horses are bred for traits deemed desirable in competition, many become less healthy and often lose the ability to breed naturally.[12] Will the same happen with humans? Will we create a class-based society, consisting of enhanced humans versus traditional ones who did not benefit from fetal enhancement, gene therapy, stem cells, and many expensive new health treatments? And if so, what will it mean to be human in such a world?

Endnotes

[1] *The Selfish Gene* is a provocative book on evolution by biologist and best-selling author Richard Dawkins, published in 1976.

[2] Ron Zemke, Claire Raines, and Bob Filipczak, *Generations at Work: Managing the Clash of Veterans, Boomers, Xers, and Nexters in Your Workplace* (New York: AMACON, 1999).

[3] Richard Foster, *Innovation: The Attacker's Advantage* (New York: Summit Books, 1986); Clay Christensen, *The Innovator's Dilemma* (Boston: Harvard Business Press, 1997).

[4] Richard Foster, *Creative Destruction* (New York: Doubleday, 2001); Clay Christensen, *The Innovator's Solution* (Boston: Harvard Business Press, 2003).

[5] George S. Day and Paul J. H. Schoemaker, eds., *Wharton on Managing Emerging Technologies* (New York: John Wiley & Sons, 2000).

[6] Paul J. H. Schoemaker, *Profiting from Uncertainty* (New York: The Free Press, 2002).

[7] Vijay Govindarajan and Chris Trimble, *10 Rules for Strategic Innovators* (Boston: Harvard Business School Press, 2005).

[8] Adam M. Brandenburger, Barry J. Nalebuff, and Ada Brandenburger, *Co-opetition* (New York: Doubleday, 1997).

[9] George S. Day and Paul J. H. Schoemaker, *Peripheral Vision* (Boston: Harvard Business School Press, 2006).

[10] Kenneth Boulding, "On Possible Consequences of Increased Life Expectancy," *Population and Development Review* 29:3 (September 2003): 493–504. Available at www.jstor.org/stable/3115288.

[11] Jeremy Rifkin's book *The Biotech Century: Harnessing the Gene and Remaking the World* (New York: Tarcher/Putnam, 1999) sounded an early warning about some troubling ethical issues. Gregory Stock's *Redesigning Humans: Choosing Our Genes, Changing Our Future* (New York: Houghton Mifflin, 2003) discusses how designer genes will let us choose the traits of our offspring and the ethical issues this raises. Lee M. Silver's *Remaking Eden: How Genetic Engineering and Cloning Will Transform the American Family* (New York: Harper Perennial, 2007) likewise explores how biotech will enable us to design our offspring and where this will lead society.

[12] Jane Stern and Michael Stern, *Dog Eat Dog: A Very Human Book About Dogs and Dog Shows* (New York: Fireside, 1998).

A

DNA, RNA, and protein

Molecules of inheritance

In the early 1900s, DNA was shown to consist of building blocks called *nucleotides* (see Figure 2.2 in Chapter 2). At the time, DNA was not considered the likely molecule of inheritance. Protein, the other known component of chromosomes, was the favored molecule. It was known that proteins consisted of amino acids; at least 20 had been discovered, whereas DNA had only 4 different nucleotides. In view of its more complex structure, protein was thought to be more capable of storing the vast amounts of information needed for millions of inherited traits than the simpler DNA structure with just four different nucleotides.

However, the protein model of inheritance had fallen into doubt by 1944, following experiments in Oswald Avery's lab at Rockefeller Institute (now Rockefeller University) in New York. Avery continued a line of experimentation started two decades earlier by British scientist Fred Griffith. Griffith worked with strains of pneumococci, which in virulent form can cause pneumonia. He observed that when two strains of innocuous bacteria—one dead but formerly virulent strain (with a smooth capsule surrounding the cell) and one live nonvirulent strain (rough appearance due to lack of capsule)—were mixed together and injected into mice, the mice got sick and died as if they had been infected by the virulent strain alone (see Figure A.1). How could combining the two form such a deadly mix, especially because the virulent strain was presumably killed? A further mystery was that the dead mice contained live, virulent (smooth capsule) bacteria.

Avery and his colleagues speculated that some factor from the dead, virulent strain had transformed the live, nonvirulent strain into a killer. They determined what that factor was by systematically destroying different parts of the virulent bacteria. By using enzymes that destroy various types of molecules—RNase for RNA, proteases to degrade proteins, and DNase, which destroys DNA—they discovered that only when material from the virulent bacteria was treated with DNase could it no longer transform non-virulent bacteria into killers. They hit the jackpot: DNA from the virulent strain must be transforming the non-virulent strain into a killer, and therefore DNA must be the bearer of genetic traits.[1]

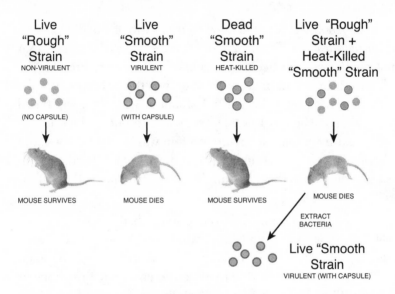

Figure A.1   Griffith's experiment

The structure of DNA

Soon afterward, in 1953, James Watson and Francis Crick, from Cambridge University, presented their theory for the three-dimensional structure for DNA, a giant milestone in science. They proposed that DNA consists of two chains that wind around each other in a helical orientation (refer to Figure 2.2 in Chapter 2) and are held together by weak bonds between hydrogen atoms located on the four nucleotides. These bonds held the strands in place but also permitted them to separate for the next round of replication.

Proof for the unzipping process of DNA replication, as suggested by Watson and Crick, came five years later with the elegant work of Meselson and Stahl (see the following sidebar).

The most beautiful experiment

Centrifuges spinning at high speed are commonly used to separate molecules in a solution according to weight, with the heavier molecules migrating faster to the bottom of the solution than the lighter molecules. Because the chemical bases of DNA contain nitrogen, Matt Meselson and Frank Stahl used an isotope of nitrogen that was heavier than normal nitrogen.[2] Bacteria were initially grown in a medium containing the heavy form of nitrogen. This nitrogen was taken up into all bacterial DNA as these cells grew and divided to form daughter cells, such that heavy nitrogen replaced normal nitrogen in all cells. Some of these "heavy" cells were then placed in a new medium containing only light (normal) nitrogen.

Meselson and Stahl argued that if Watson and Crick were correct, a hybrid DNA molecule would form from the first round of replication, with the parental strand of the newly formed DNA containing heavy nitrogen and the new daughter strand containing the light form. This hybrid molecule would be halfway between the weight of the original heavy parental DNA and the all-light daughter DNA, and could therefore be separated in the centrifuge. Indeed, the centrifugation showed three distinct bands of DNA in the test tubes, confirming the existence of the hybrid molecule and supporting Watson and Crick's model of DNA replication.[3]

Cracking the code

After the structure of DNA and a model for how it replicates had been laid out, the next big milestone was to crack the genetic code—that is, explain how DNA chains containing just four different nucleotides could direct the synthesis of all cellular proteins. Proteins are the workhorses of the cell, catalyzing its myriad biochemical reactions and making up a good part of the cell's structural components, such as organelles and membranes.

How could the four different nucleotides in DNA possibly code for the 20 different amino acids found in proteins? Clearly, the genetic code must have more than one or even two nucleotides: Even a 2-nucleotide code would, at most, be able to specify 16 different amino acids (4 nucleotides combined 2 at a time)—not enough, given the 20 known amino acids. The code would have to have a minimum of three nucleotides, which would allow 64 distinct combinations. A flurry of experiments in the 1960s finally "cracked" the full code, determining the combination of 3 nucleotides that specified each of the 20 amino acids.

Making proteins

Many experiments over several years uncovered the fine details of protein synthesis. Several different types of RNA are involved (see the following sidebar). To make a cellular protein, such as insulin, the two strands of DNA in the cell nucleus separate, or unzip. With the help of an enzyme, RNA polymerase, and using one of the two strands as a template, nucleotides from the surrounding cytoplasm assemble and are linked together to form messenger RNA (mRNA).

What is RNA?

RNA differs from DNA, in that it occurs as a single chain composed of nucleotides, whereas DNA is a double chain. Also, RNA's nucleotides have either an adenine, guanine, cytosine, or uracil nitrogen base, whereas DNA contains thymine instead of uracil. RNA also contains a slightly different form of sugar, ribose, than DNA, which contains deoxyribose. To study the role of various RNAs in protein synthesis, cell-free experiments in test tubes were conducted by adding various components of the cell. They showed that ribosomes, distinct RNA- and protein-containing structures, were the site of protein assembly. Also, it was shown that before amino acids were incorporated into a protein chain, they were attached to small RNA molecules, termed transfer RNA. Each amino acid had its own specific transfer RNA. A third type of RNA, dubbed messenger RNA, is the actual template for protein synthesis. Several laboratories, including James Watson's at Harvard, elucidated the mechanics of protein assembly: Ribosomes provided the cellular venue for mRNA strands to direct the synthesis of new proteins using amino acids ferried to the ribosomes by transfer RNA.

The process is called *transcription*. Messenger RNA then goes to the ribosomes in the cytoplasm, where its message—a sequence of triplet codons specific for a protein—is translated into the sequence of amino acids that make up that protein. Transfer RNAs are the linchpin, pulling amino acids from the cytoplasm and tugging them to the mRNA, where they can be assembled into protein (see Figure A.2).

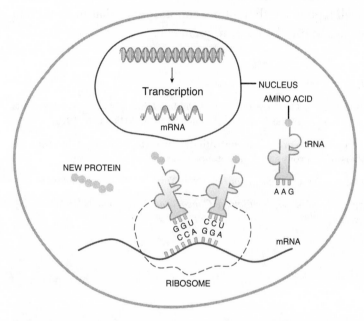

Figure A.2 Protein synthesis

Regulating gene expression

The next discovery in the genetic story was figuring out how protein synthesis was controlled. If every cell in the human body contains identical DNA, how can cells in different tissues be so markedly different? Insights into this mechanism came in the 1960s, when Francois Jacob and Jacques Monod described genes switching on and off in the laboratory bacterium *Escherichia coli*. When the bacterium was grown in a broth containing the sugar lactose, it produced an enzyme (beta-galactosidase) needed to break down this sugar for energy. Without lactose in the medium, no such enzyme was produced. Jacob and Monod postulated that a repressor molecule normally prevents the transcription of the beta-galactosidase gene and that this repressor is inactivated by lactose. Over the years, similar methods for turning genes on and off were shown to occur in higher organisms, including humans, along with additional regulatory processes not found in bacteria[4]. Regulation of gene expression gives an organism the ability to conserve energy because the genes are switched on only when their protein products are needed. This is clearly an evolutionary advantage. It also accounts for differences from one cell type to another—for example, genes that make a brain cell unique are transcribed only in brain cells.

Endnotes

[1]F. H. Portugal and J. S. Cohen, *A Century of DNA: A History of the Discovery of the Structure and Function of the Genetic Substance* (Boston: MIT Press, 1979).

[2]An isotope is one of two or more atoms with the same atomic number, but with different numbers of neutrons and different molecular weights.

[3]James D. Watson, *DNA: The Secret of Life* (New York: Alfred A. Knopf, 2003).

[4]Gerald Karp, *Cell and Molecular Biology: Concepts and Experiments* (Hoboken, NJ: John Wiley & Sons, 2008).

B

Cloning genes

We describe a typical cloning procedure, starting with an illustration of how restriction enzymes work in the process.

Restriction enzymes

One of the first restriction enzymes, dubbed EcoR1, was discovered in E. coli and found to cut only within the nucleotide sequence GAATTC (or its mirror image CTTAAG), and specifically between the G and A. The enzyme cuts both DNA strands of the double helix creating small single-stranded fragments at the cut sites that stick out:

Double-stranded DNA before cutting:

XXXXXXXGAATTCXXXXXXX
XXXXXXXCTTAAGXXXXXXX

Double-stranded DNA after cutting with EcoR1:

XXXXXXXG   AATTCXXXXXXX
XXXXXXXCTTAA GXXXXXXX

Note that the Xs in this example can be A, T, C, or G, corresponding to the basic alphabet of the genetic code. The dangling TTAA and AATT ends are termed *sticky ends* because they can come back together, as in the top illustration, by complimentary base pairing (A with T and C with G) and, thus, create an intact DNA molecule. These dangling ends can also facilitate the joining of two distinct DNA molecules that have been cut with the same restriction enzyme.[1]

For example, a fragment of DNA containing the insulin gene from humans and a bacterial plasmid could be cut with the same restriction enzyme. The complementary sticky ends that result will allow the human gene to attach to the bacterial plasmid. This recombinant plasmid can then be used to clone multiple copies of the insulin gene.

In addition to EcoR1, over a thousand restriction enzymes have been identified in a range of different bacteria, each with different cutting specificities. It quickly became evident that these would be valuable tools for precisely cutting any DNA, even that from human cells, because all DNA molecules, no matter the creature, are made up of the same four nucleotides. The beauty of these enzymes is their specificity; they each cut at a specific sequence of nucleotides reliably in each experiment and, thus, can be used as precision tools for making DNAs of manageable lengths for further study.

Cloning procedure

In a typical cloning experiment (see Figure B.1), a target gene, such as the insulin gene, is isolated from its source using a restriction enzyme. The fragment of DNA containing the target gene is then joined to a plasmid DNA whose circular strand was cut with the same restriction enzyme, so that both DNAs have complementary "sticky" ends. The plasmid DNA codes for resistance to a specific antibiotic, such as penicillin. The target DNA and plasmid DNA are joined together by ligase, another enzyme in the "cut and paste" biotech toolkit. Several types of molecules can result from the cutting and pasting process: Linear strands can recircularize to form the original plasmids, or novel hybrid plasmids can form, containing DNA fragments from both the plasmid and target DNA.

The recombinant plasmids are then mixed with bacteria in a test tube. These bacteria are induced to take up the recombinant plasmid into their cytoplasm, a process called transformation. To select bacteria containing recombinant plasmids, bacterial cells are plated on growth media containing an antibiotic; those that grow contain plasmids with resistance genes. Some of those bacteria will also contain the insulin gene. Once inside bacterial cells, recombinant plasmids will replicate into hundreds of copies, thus providing amplification of the gene of interest.

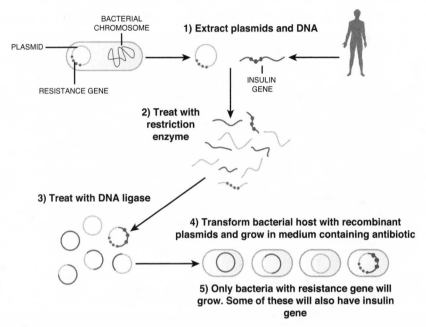

Figure B.1 Typical cloning procedure

Endnotes

[1]Reinhard Renneberg, *Biotechnology for Beginners* (Berlin: Springer-Verlag, 2008).

C

Complexity of the genome

The total number of human genes identified by from the Human Genome Project—approximately 25,000—is much less than originally expected, considering the complexity of humans. However, it appears that most of our mRNA can be spliced (cut) and reassembled in alternate ways, giving rise to many more mRNAs and proteins than are encoded in our genome. Scientists believe that up to 60% of human genes can undergo alternative splicing of their mRNAs.[1] After mRNA is transcribed from DNA in the genome, introns are cut out (a process called splicing), leaving only intact coding sequences called exons. In addition to splicing out introns, the remaining mRNA can undergo further splicing. The sequences that are spliced can differ depending on which tissue the mRNA is located in and what job the resulting protein must perform.

Non-coding RNAs

As much as 98% of the human genome does not code for proteins. This so-called "junk DNA" was thought to have no function, but recent research shows that virtually all of the cell's DNA is copied into RNA. What does all this RNA actually do in the cell? In the past several years, various classes of noncoding RNAs (as opposed to messenger RNAs that code for proteins) have been discovered. One of the most actively researched noncoding RNAs are the microRNAs (miRNAs). These are extremely small, containing only 21–25 nucleotides, and were originally overlooked. However, researchers are increasingly realizing their importance in regulating gene expression in both the animal and plant kingdoms. MicroRNAs play critical regulatory roles in embryo development, as well as in the expression of genes in various tissues, as needed. By binding to mRNA at regions of homology, miRNAs can block the synthesis of protein from the mRNA, leading to less of that protein in the cell at critical times in development.[2]

At least 500 different miRNAs have so far been discovered in humans, and each of these can interfere with the activity of many different mRNAs. MicroRNAs are involved in many disease processes, such as cancer, diabetes, and heart failure. In one study, mice were engineered to lack a specific miRNA, which is normally highly expressed in heart muscle. Without that miRNA, half of the animals had holes in their hearts or fatal cardiac rhythm disruptions.[3] Several biotech companies are focusing on miRNAs and their potential to diagnose or treat diseases. In addition to the miRNAs, researchers have discovered several other classes of non-protein-coding RNAs (siRNAs and Piwi-interacting RNAs) in plants and animals, and are investigating their regulatory functions.[4]

SNP variations

SNPs stand for *single nucleotide polymorphisms.* These are variations in the DNA sequence that involve an alteration in one nucleotide base pair (A,T,C, or G). SNPs have arisen over centuries of human evolution and, therefore, are shared by human subpopulations. For example, in about every 1000 base pairs, two random individuals will exhibit a difference in their DNA sequence—one person might have a G at that location, while the other person has a T. This adds up to millions of SNP differences between humans.

Because about 99.5% of human DNA sequences are the same, these variations have diagnostic value.[5] SNPs are of interest for medical as well as pharmacological research. Researchers can now scan the genomes of multiple individuals with the same disease to look for SNPs associated with that disease that healthy individuals do not have. If they find such an SNP, it could contribute to the disease process or act as a marker for a nearby mutation that is involved. In addition, profiling a person's SNPs could tell researchers whether certain drugs will be more or less likely to work in that individual. This will expand the era of customized (personalized) drug therapy.

Although SNPs account for about 90% of human genetic variation, genomic studies have revealed other types of variation as well. These include structural differences in the DNA sequence, such as extra nucleotides, deleted nucleotides, repeated sequences, and inversions of the DNA sequence. All of these contribute to variability within the human population, both in disease susceptibility and in traits unrelated to disease.

Endnotes

[1]Gerald Karp, *Cell and Molecular Biology, Concepts and Experiments* (Hoboken, NJ: John Wiley & Sons, 2008).

[2]Tina Hesman Saey, "Micromanagers: New classes of RNAs emerge as key players in the brain," *Science News* 173 (2008): 136.

[3]Jennifer Couzin, "MicroRNAs Make Big Impression in Disease After Disease," *Science* 319 (2008): 1782.

[4]James A. Birchler and Harsh H. Kavi, "Slicing and Dicing for Small RNAs," *Science* 320 (2008): 1023.

[5]Jon Cohen, "Venter's Genome Sheds New Light on Human Variation," *Science* 317 (2007): 1311.

Glossary of Biomedical Terms

amino acid   An organic molecule containing three functional groups—an amino group, a carboxyl group, and a distinct side chain—attached to a central carbon. Amino acids are the building blocks of proteins.

angiogenesis   The formation of new blood vessels, for example, to supply blood to a new tumor.

antibody   A protein derived from B lymphocytes that attaches to surface molecules on bacteria, viruses, or cancer cells and facilitates their destruction by the immune system.

antigen   Any molecule or substance that the immune system recognizes as foreign.

antisense therapy   When the genetic sequence of a target gene involved in disease is known, scientists can construct a complementary RNA sequence (antisense RNA) that binds to the mRNA from that gene (sense RNA), creating a double-stranded molecule. This effectively turns off the target gene because mRNA must be single-stranded to be translated into protein.

base pair   Two nitrogen-containing molecules (adenine, cytosine, guanine, or thymine) that are paired together in DNA and are held close by a weak bond; adenine pairs with thymine and cytosine pairs with guanine. This binding between complementary base pairs holds the double helix together. Nitrogen bases are one of the components of nucleotides, the building blocks of DNA and RNA.

biochip   Any device that incorporates biological or organic molecules on a solid surface to perform biochemical reactions; examples are DNA and protein microarrays and glucose sensors to measure blood glucose. They have some similarity to integrated circuits (microchips) in their manufacturing.

bioinformatics   The application of information technology to the field of molecular biology and medicine. This includes computationally intensive applications such as mapping and analyzing DNA and protein sequences.

biomarker   A substance used as an indicator of a biological state, most commonly disease.

biometrics   Use of human body characteristics such as fingerprints, eye retinas and irises, voice patterns, facial patterns, and hand measurements for authentication purposes.

biomimetics   The application of methods and systems found in nature, to engineering and biotechnology.

chromosomes   Strands of DNA that carry the genetic information of the human (or animal) genome; they can be seen in the nucleus just before cell division. Human cells contain 23 chromosome pairs; the father and mother each contribute 1 member of the pair.

cloning   The replication of a DNA sequence—more generally, the production of a cell or organism with the same DNA as another cell or organism.

codon   Three contiguous nucleotides in messenger RNA that specify a particular amino acid.

complementary base pairing   The relationship between the nitrogen base components (adenine, thymine, cytosine, guanine) of nucleotides that are opposite each other in the two strands of the double helix of DNA. For structural reasons, only adenine can pair up opposite thymine, and cytosine with guanine.

cytoplasm   The content or material inside a cell, excluding the nucleus.

DNA (deoxyribonucleic acid)   A molecule carrying genetic information that consists of two chains of nucleotides configured in a double helix. DNA carries the genetic information used for the development and functioning of organisms.

DNA sequencing A method for determining the order of nucleotides in a given strand of DNA.

enzyme A molecule (generally a protein) that catalyzes or increases the rate of chemical reactions. The enzyme RNA polymerase, for example, is involved in RNA synthesis.

eugenics The study of genetically improving the human race by using controlled selective breeding.

exon The sequence of nucleotides within a split gene that codes for a specific protein.

free radical A highly reactive atom or molecule that can damage DNA or other cellular structures.

gene A segment of the DNA that occupies a specific location on the chromosome and determines a particular characteristic of an organism.

gene expression The process by which information contained in a gene is used to create gene products, such as RNA or proteins. Although every human cell contains the same DNA, the expression of particular genes allows one type of cell to differ from another in structure and function.

gene therapy Treating a disease by changing the nucleotide sequence of a specific gene that causes the disease.

genetic code The code that translates nucleic acid sequence into protein sequence. More specifically, it is the relationship between the nucleotide base-pair triplets of a messenger RNA molecule and the 20 amino acids that are the building blocks of proteins.

genetically modified organism (GMO) An organism whose genetic material has been altered using genetic engineering techniques.

genome The entire set of genes within an organism's chromosomes.

genomics   The study of the genetic content of an organism (its genome).

hormone   A chemical messenger released from one cell that carries a signal to another cell to initiate a cellular response.

hybridoma   The hybrid cell resulting from the fusion of a malignant cancer cell with an antibody-producing lymphocyte cell. This hybrid cell continuously divides in culture to produce a clone of cells, yielding large amounts of the single antibody (monoclonal antibody) produced by the original lymphocyte.

immunoassays   Biochemical tests that measure the concentration of a substance in a biological fluid such as blood or urine by measuring the degree of reaction of an antibody specific to the substance. For example, the substance could be a hormone, such as insulin, or an infectious agent, such as the HIV virus that causes AIDS.

introns   The sequences of nucleotides within a split gene that lie between coding sequences (exons) and that do not code for protein; they are also called intervening sequences.

isotope   One of two or more atoms with the same number of protons in the nucleus but a different number of neutrons. Carbon 12 and Carbon 14 are both isotopes of carbon, one with six neutrons and one with eight neutrons.

longevity   Length or duration of life; also long duration of life.

lymphocyte   A white blood cell that is one of two forms—either T or B—that circulates between the lymph and blood systems and is involved in the immune response.

MEMS (micro-electrical-mechanical systems) technology   Micro applications of biomedical engineering, such as small chips used as implantable replacements for diseased organs in the body.

monoclonal antibody   A single type of antibody produced by a clone of identical cells.

nanotechnology   The development of materials or devices whose size is on the scale of 100 nanometers or smaller.

nucleotides   Compounds that are the building blocks of DNA and RNA. They are composed of a sugar, a phosphate group, and a nitrogen base.

organic compounds   Natural or synthetic substances whose molecules contain carbon.

personalized medicine   Medical care tailored to individuals based on their genetic makeup.

pharmacodynamics   Tracking illness-related changes in the body, such as those induced by a patient's drug-treatment protocol.

pharmacogenetics   Prescribing the type and dosage of medicine based on a patient's unique genetic profile.

plasmid   Self-replicating circular DNA molecules found in bacteria. Plasmids can be used to replicate foreign DNA inserted into the plasmid.

pluripotent   Having the capacity to develop into all the cell types in the human body, as embryonic stem cells do.

polymerase chain reaction (PCR)   A method for greatly amplifying the number of copies of a single fragment of DNA that is originally present in very small amounts.

prolongevity   The process of extending the life span of an individual or population through interventions that promote better use of preventive medicine and use of established diagnostic and therapeutic facilities.

proteins   Complex macromolecules made up of long chains of amino acids that play a wide range of roles in cells, from making up structural components to being involved in biochemical reactions as enzymes.

proteomics   Involves the identification of proteins in an organism and the determination of their role in normal physiology as well as disease.

recombinant DNA   Novel DNA molecules constructed from DNA fragments from different sources.

regenerative medicine   Process of creating living, functioning replacement tissues or organs for tissues or organs that have lost function from disease, aging, or damage.

reproductive cloning   Producing a clone of animal X by fusing one of its body cells with an egg cell from donor Y that has had its nucleus removed. This fusion cell divides to become an adult that has the genetic traits of animal X.

restriction enzymes   Enzymes from bacteria that recognize specific nucleotide sequences in DNA and cut specifically at those sites on both strands of the double helix.

ribosomes   Cellular structures composed of RNA and proteins that serve as sites for protein synthesis.

RNA (ribonucleic acid)   A chain of nucleotides that is similar to DNA, except that it contains the nitrogen base uracil instead of thymine, contains the sugar ribose instead of deoxyribose, and is usually single-stranded. RNA is involved in translating the genetic code in DNA into proteins.

RNA interference (RNAi)   The process whereby short, double-stranded RNAs, which are complementary to target mRNAs, can bind to and direct the degradation of the target mRNAs. This prevents production of the proteins encoded by the mRNAs.

single nucleotide polymorphism   A difference among individuals in one nucleotide in a sequence of DNA.

somatic cell nuclear transfer   Transferring the nucleus from a body cell of an animal or human donor into the cytoplasm of an enucleated egg (nucleus removed) from a second donor. The newly created embryo can be used to harvest stem cells that have the genetic traits of the nucleus donor, or can be allowed to grow into an adult that is a clone of the nucleus donor.

stem cells   Undifferentiated cells from the embryo that can develop into all the various cell types in the adult animal or human (embryonic stem cells), or cells located in various body tissues that will replace all the various types of cells in that particular tissue as needed (adult stem cells).

synthetic chemistry   A branch of chemistry in which chemists devise ways to make specific compounds of interest and/or develop new chemical reactions for this purpose.

systems biology   An interdisciplinary approach to biology that focuses on the systematic study of complex interactions in biological systems, viewing the systems as a whole instead of reducing them to their individual parts.

telemedicine   Medical information that is conveyed from one site to another via electronic communications, for the purpose of improving patient care. It includes consultative, diagnostic, and treatment services.

therapeutic cloning   Producing an embryo by fusing a body cell from a human patient X with an egg cell from a donor that has had its nucleus removed. The fusion cell divides and becomes an embryo that has the genetic traits of patient X. Stem cells from the embryo can be used as therapies for that patient, such as when treating diabetes or Parkinson's disease.

tissue engineering   Applies principles of engineering and life sciences toward the goal of developing biological substitutes to restore, maintain, or improve tissue or organ function.

transcription   The process by which RNA is synthesized from DNA.

translation   Transfer of genetic information coded in messenger RNA into the manufacture of proteins.

virulence   The severity of disease caused by a microorganism, in terms of the symptoms it causes or the ease of spread through a population.

xenotransplantation   Transplantation of cells, tissues, or organs from animals into humans.

Acknowledgments

Our book builds on a biosciences research study we conducted with the Mack Center for Technological Innovation, at the Wharton School of the University of Pennsylvania. We deeply thank the Center's codirectors, Professors George Day and Harbir Singh, for their unwavering financial and intellectual support, and also Michael Tomczyk, the Center's managing director, for his help as coeditor of the original report. The Wharton study was partly funded by Akzo-Organon, Deloitte Consulting, Hewlett-Packard, Hitachi Chemical Research Center Inc., Infosys Technologies, Johnson and Johnson, Procter and Gamble, Reliance Group, Siemens Corporation, Strategic Decisions Group (SDG), and Quest Diagnostics, who are all gratefully acknowledged. We also thank BIO, the world's largest biotechnology industry organization, for the invitation to present our findings at their annual conference.

In addition, we thank Decision Strategies International, the consultancy that helped conduct the initial Wharton study, for its kind permission to draw on the joint report. Specifically, we thank Jim Austin, Michael Mavaddat, and Scott Snyder for their intellectual contributions, as well as Nanda Ramanujan. We also benefited from the able research assistance of Kunal Bahl, Gary Kurnov, Ari Meisel, and Pranav Vora for the original study and especially the excellent background research of David Huang for the present book. In addition, we received valuable feedback on our book manuscript from Jim Austin, Lawton Burns, Maureen Evans, Josh Hyatt, Kirk Jensen, Donald Kalff, David Lester, Marianne Rosemann, Leanne Wagner, and Charity Vitale as well as various members of the Pearson team. We especially thank Professor Jerry Wind from Wharton for suggesting this book idea to Pearson Publishing in the first place, and Tim Moore for pursuing the idea further as senior editor.

We also wish to thank the various commentators, conference presenters, and facilitators associated with the original Wharton project. They include Pradip Banerjee, Lawton (Rob) Burns, Arthur Caplan, Tom Donaldson, Terry Fadem, Richard Foster, William Hamilton, Robert Lindenschmidt, George Milne, Stelio Papadopolous, Stephen Sammut,

Kiran Mazumdar-Shaw, Nicolaj Siggelkow, K. V. Subramaniam, David U'Prichard, and Sidney Winter. We especially thank Larry Huston who, while still at P&G, encouraged us several years ago to study the biosciences field in greater detail as part of the Mack Center's research agenda. He helped conceive the initial research agenda and generously shared his industry maps. In addition, numerous experts contributed to our research through their participation in interviews as well as various workshops we convened at the Wharton School. Some of these experts, but by no means all, are quoted in this book.

Specifically, we wish to acknowledge Neha Amin, Raffi Amit, Howard Bilofsky, Frank Bianchetti, Fritz Bittenbender, Michael Breggar, Patrick Caubel, Sriram Chandrasekaran, Paul Citron, Floyd Cole, Christine Côté, Hugh Courtney, Patricia Danzon, Mark Day, Nadina Deigh, Don Doering, Jeff Edelson, Irv Epstein, J. William Freytag, Lance Gordon, Eileen Gorman, Jim Greenwood, Hank Guckes, Steve Haeckel, Chris Harley-Martin, Warren Haug, Witold Henisz, Terry Hisey, Denise Hudson, Heidi Hunter, Robert Johansen, Munsok Kim, Ronald Krall, Jeff Lerner, Marie Lindner, Shawn Lufstrom, Kevin Mansmann, Patia McGrath, Marcus McNabb, Edward Miller, Graham Mitchell, Mark Myers, Scott Nichols, Marvin Ng, Sharon Nunes, Anil Patel, Jerry Parrot, Basab Pradhan, Sandeep Raju, Sandy Retzky, Robert C. Riley, Barry Robson, Ron Ruggieri, Bill Semancik, Ari Sherwood, Megan Simmonds-Luce, Robert Smith, Ron Taylor, Thomas Tillett, William Tulskie, Dong Wei, and James Wilson.

About the Authors

Paul J. H. Schoemaker has worked for many years as a professor in academia (at University of Chicago and the Wharton School of the University of Pennsylvania) and also as a leader in business (with Royal Dutch/Shell and Decision Strategies International). This book builds on his recent work at the Wharton School, where he serves as Research Director of the Mack Center for Technological Innovation. Paul was a co-founder of various new ventures and has been a consultant and speaker to companies around the world. He is the coauthor of eight books, including *Decision Traps* (Doubleday, 1989), *Wharton on Emerging Technologies* (Wiley, 2000), *Winning Decisions* (Doubleday, 2001), *Profiting from Uncertainty* (Free Press, 2002) and *Peripheral Vision* (Harvard Business School Press, 2006), and has written more than a hundred articles. He earned his undergraduate degree in physics, and he has an MBA and Ph.D. from the Wharton School. Paul was born and raised in the Netherlands.

Joyce A. Schoemaker is a scientist with a doctorate in microbiology from Thomas Jefferson University in Philadelphia. She held a postdoctoral appointment at the University of Chicago, where she published various scientific articles. Joyce was a senior scientist in London at Celltech Ltd., England's first biotech company, and served as Director of Molecular Biology at BW Biotech, a Chicago-based biotechnology company. She has taught biology courses at St. Josephs College and Villanova University, and is the coauthor of *Healthy Homes, Healthy Kids: Protecting Your Children from Everyday Environmental Hazards* (Island Press, 1991). Joyce has a long-standing interest in environmental issues and the emerging biosciences.

Index

A

Abbott Labs, 88, 93-94
access to medical data, 74-75
acquired immune deficiency syndrome. *See* AIDS
adult stem cells, 51
aging process, potential methods for slowing, 4-7
agricultural industry
  genetically modified crops, 63
  recombinant DNA technology in, 30
AIDS (acquired immune deficiency syndrome), 109-111, 130
  treatment options, 115
Amgen, 88
amino acids, 26
  role in protein synthesis, 179-180
anatomic imaging, 61
angiogenesis, 72
angiogenesis inhibitors, 119
angiogenesis therapy, 115
angioplasty, 115
Animal Liberation Front, 129
animal rights organizations, 128-129
anthrax attacks, 131
antibiotics, 25-26, 51
  resistance to, 28, 52
antibodies, monoclonal, 24, 49
antigen vaccines, 118
antigens, 47
antimicrobials, 50-51
antiretroviral drugs, 110
antisense technology, 45-46

antivirals, 51
apathy in bio gridlock scenario, 150
Apple, 162
applied bioscience, disciplines in, 35
Aranesp, 88
arrested aging, 7
artemisinin, 59
arthroscopic knee surgery, 6
artificial knee joints, 6
assisted living, resources for, 12
attenuated vaccines, 47
Aum Shinrikyo, 131
Avastin, 49
Avery, Oswald, 177
avian flu, 129

B

B lymphocytes, 24
bacteria, cloning genes in, 34n
balloon angioplasty, 115
Baltimore, David, 34n
basic bioscience, disciplines in, 35
Basic Local Alignment Search Tool (BLAST), 71
Bayh, Birch, 19n
Bayh–Dole Act of 1980, 13
Beckman Coulter, 94
bio gridlock scenario, 140, 143-153
  analogous cases, 151-153
  apathy in, 150
  causes of, 145-150
  consumer representation in, 151
  economic impact in, 149-150

Page numbers followed by *n* indicate topics located in endnotes.

204

Index

FINANCIAL TIMES

In an increasingly competitive world, it is quality
of thinking that gives an edge—an idea that opens new
doors, a technique that solves a problem, or an insight
that simply helps make sense of it all.

We work with leading authors in the various arenas
of business and finance to bring cutting-edge thinking
and best-learning practices to a global market.

It is our goal to create world-class print publications
and electronic products that give readers
knowledge and understanding that can then be
applied, whether studying or at work.

To find out more about our business
products, you can visit us at www.ftpress.com.